NOSTALGIC VIRILITY AS A CAUSE OF WAR

McGill-Queen's/Brian Mulroney Institute of Government Studies in Leadership, Public Policy, and Governance

SERIES EDITOR: DONALD E. ABELSON

Titles in this series address critical issues facing Canada at home and abroad and the efforts policymakers at all levels of government have made to address a host of complex and multifaceted policy concerns. Books in this series receive financial support from the Brian Mulroney Institute of Government at St Francis Xavier University; in keeping with the institute's mandate, these studies explore how leaders involved in key policy initiatives arrived at their decisions and what lessons can be learned. Combining rigorous academic analysis with thoughtful recommendations, this series compels readers to think more critically about how and why elected officials make certain policy choices, and how, in concert with other stakeholders, they can better navigate an increasingly complicated and crowded marketplace of ideas.

5 Government
Have Presidents and Prime Ministers Misdiagnosed the Patient?
Donald J. Savoie

6 Cyber-Threats to Canadian Democracy
Edited by Holly Ann Garnett and Michael Pal

7 The Canadian Federal Election of 2021
Edited by Jon H. Pammett and Christopher Dornan

8 CETA Implementation and Implications
Unravelling the Puzzle
Edited by Robert G. Finbow

9 Multilateral Sanctions Revisited
Lessons Learned from Margaret Doxey
Edited by Andrea Charron and Clara Portela

10 Booze, Cigarettes, and Constitutional Dust-Ups
Canada's Quest for Interprovincial Free Trade
Ryan Manucha

11 NORAD
In Perpetuity and Beyond
Andrea Charron and James Fergusson

12 Under the Weather
Reimagining Mobility in the Climate Crisis
Stephanie Sodero

13 Rethinking Decentralization
Mapping the Meaning of Subsidiarity in Federal Political Culture
Jacob Deem

14 Natural Allies
Environment, Energy, and the History of US-Canada Relations
Daniel Macfarlane

15 Gender, Communications, and Reproductive Health in International Development
Carolina Matos

16 Nostalgic Virility as a Cause of War
How Leaders of Great Powers Cope with Status Decline
Matthieu Grandpierron

NOSTALGIC VIRILITY AS A CAUSE OF WAR

How Leaders of Great Powers Cope with Status Decline

MATTHIEU GRANDPIERRON

McGill-Queen's University Press
Montreal & Kingston | London | Chicago

ISBN 978-0-2280-2016-5 (cloth)
ISBN 978-0-2280-2036-3 (paper)
ISBN 978-0-2280-2017-2 (ePDF)
ISBN 978-0-2280-2018-9 (ePUB)

Legal deposit first quarter 2024
Bibliothèque nationale du Québec

Printed in Canada on acid-free paper that is 100% ancient forest free (100% post-consumer recycled), processed chlorine free

Funded by the Government of Canada | Financé par le gouvernement du Canada | Canada

Canada Council for the Arts | Conseil des arts du Canada

We acknowledge the support of the Canada Council for the Arts.

Nous remercions le Conseil des arts du Canada de son soutien.

McGill-Queen's University Press in Montreal is on land which long served as a site of meeting and exchange amongst Indigenous Peoples, including the Haudenosaunee and Anishinabeg nations. In Kingston it is situated on the territory of the Haudenosaunee and Anishinaabek. We acknowledge and thank the diverse Indigenous Peoples whose footsteps have marked these territories on which peoples of the world now gather.

LIBRARY AND ARCHIVES CANADA CATALOGUING IN PUBLICATION

Title: Nostalgic virility as a cause of war : how leaders of great powers cope with status decline / Matthieu Grandpierron.
Names: Grandpierron, Matthieu, author.
Series: McGill-Queen's/Brian Mulroney Institute of Government studies in leadership, public policy, and governance ; 16.
Description: Series statement: McGill-Queen's/Brian Mulroney Institute of Government studies in leadership, public policy, and governance ; 16 | Includes bibliographical references and index.
Identifiers: Canadiana (print) 20230501052 | Canadiana (ebook) 20230501117 | ISBN 9780228020165 (cloth) | ISBN 9780228020363 (paper) | ISBN 9780228020189 (ePUB) | ISBN 9780228020172 (ePDF)
Subjects: LCSH: Great powers. | LCSH: International relations. | LCSH: Power (Social sciences) | LCSH: Indochinese War, 1946-1954—Causes. | LCSH: Falkland Islands War, 1982—Causes. | LCSH: Grenada—History—American Invasion, 1983—Causes. | LCSH: France—Military policy. | LCSH: Great Britain—Military policy. | LCSH: United States—Military policy.
Classification: LCC JZ1310 .G73 2024 | DDC 327.101—dc23

Set in 11.5/14 Bulmer MT Standard with Ordax
Book design & typesetting by Garet Markvoort, zijn digital

CONTENTS

TABLES AND FIGURE

Tables

Figure

ACKNOWLEDGMENTS

To my partner Jin, without whom this project would not have been possible.

I owe a lot to Paul Saurette and Thomas Lindemann, my thesis directors, for their support and advice throughout this project.

As this book is written in English, a language that is not my mother tongue, I owe a lot to Robert Marinov, my editor, who spent hours checking the manuscript. Despite having been read and edited many times, the book may still contain some errors, for which I apologize.

NOSTALGIC VIRILITY AS A CAUSE OF WAR

INTRODUCTION

Why go to war? This has probably been the most enduring question throughout the history of mainstream, constructivist, and critical international relations (IR) literatures. Since Thucydides addressed it long ago in his *History of the Peloponnesian War*, much has been written about the causes of war – to such an extent that the question itself has become a classic. It has been investigated and answered in various ways by IR scholars, historians, lawyers, and policy-makers alike, with a common (but surely not exhaustive) list of answers including: the search for material resources; safeguarding or acquiring strategic interests; increasing national security; pursuing decision-makers' personal and/or political interests; avenging national and/or decision-makers' personal humiliations; signalling, recouping, or gaining status – and the list could go on and on.

Arguably, the question has become a classic because, in most studies, it is linked to a specific category of actor: great powers. The conflicts that garner the most attention from historians, scholars, and the public at large tend to be the major wars between great powers.[1] By studying these confrontations, attempts are made to identify lessons for the future or to predict how relations between current great powers may or may not evolve – such as, for instance, the future of US-China relations.[2]

So, if such a wealth of responses has been written on the question, why continue studying it? There are a number of compelling reasons to re-engage this age-old question. While we already know a great deal about wars between great powers, these conflicts are not only becoming rarer,[3] but war itself is changing, too. New means of conducting both major and minor wars have emerged, while the relations between war, decision-makers, and the public have also changed.[4] Even the participants in war have changed: lesser powers, "small" states, and non-state actors increasingly participate in conflicts,[5] and cases of less powerful actors challenging great powers and eventually winning have provided important new insights and theoretical developments to the literature.[6]

These changing dynamics have not only problematized many traditional assumptions, but also reinvigorated the classical question of the causes of war.

The changing nature of war is not the only compelling reason to continue studying its causes, however. While there has already been significant analysis of the two types of conflict outlined above – those between great powers and those initiated by smaller powers against great powers – there is surprisingly little academic focus on wars occurring between great powers and lesser powers *when initiated by the former*. Perhaps this relative lack of focus can be explained by what seems to be an intuitive or self-evident answer: great powers go to war against lesser powers simply because they can – they have the capacity and the advantage. After all, as Thucydides highlighted, "those in positions of power do what their power permits, while the weak have no choice but to accept it."[7] This seems an insufficient ground for neglecting research on these conflicts; yet surprisingly, there are nearly no studies focused directly on wars initiated by great powers against less powerful actors within their periphery.[8]

Despite a general lack of focus on these conflicts, there are examples throughout history of great powers engaging in strategically inconsequential regions, which highlights the question: *What are they doing there?* What makes some seemingly unattractive locations (from a strategic and/or economic point of view) suddenly worth fighting for, despite the risks and costs involved? Even a brief historical overview demonstrates that this type of war is relatively frequent. Examples include the British versus the Afghan tribes (1878), the Anglo-Zanzibar War (1896), Austria-Hungary against Serbia (1914), the first years of the First Indochina War between France and the Viet Minh (1945–47), the USSR invasion of Afghanistan (1979), the British reconquest of the Falklands (1982), the US invasions of Grenada (1983) and Panama (1989), and Russia against Georgia (2008) – not to mention the examples from antiquity, such as the final Roman destruction of Cartago during the Third Punic War (149–6 BC), or the Byzantine war of restoration in North Africa against the Vandals (533–4), and in Italy against the Ostrogoths (535–53). As these various wars were located within the great powers' peripheries, core interests were, by definition, not threatened. Options other than going to war might have been more appropriate, not to mention less risky and costly (materially and/or symbolically).

Decision-makers of great powers who have chosen to respond with the full force of their country's military were not, it would seem, forced to do so. Yet softer options, while available, were disregarded.

While aspects of great powers' interventions within their periphery have changed over time, such decisions are still relevant today and should be studied accordingly: "Strong states continue to intervene in weaker states on a massive scale when it suits them – what has changed is when it will suit them – not the fact of intervention but its form and meaning."[9] These situations and their changing dynamics will continue to confront decision-makers and IR practitioners, thus further revitalizing the age-old question of why actors choose to go to war. Although much research will be required to fully understand these types of conflicts and the reasons for their escalation into full-blown warfare, this study seeks to begin sketching out an answer.

Why This Book?

This book responds to several crucial and enduring questions in IR and strategic studies: Why do great powers go to war in their periphery? Why would decision-makers engage their country, when perceived to be in decline, in potentially long-lasting and costly conflicts? And why are softer, non-violent, or diplomatic options not prioritized in these situations? It responds to these questions by offering two main contributions to the academic literature:

By developing a novel theoretical framework centred on the study of decision-makers' status-seeking behaviours, which can help researchers, historians, and policymakers make sense of decision-makers' choices to go to war by accounting for the shortcomings of dominant IR theories, namely the realist and diversionary theories of war; and

By operationalizing this theoretical framework into a concrete set of replicable methods for empirical analysis (methods which, as discussed in the conclusion, maintain relevance for the study of both past and contemporary conflicts of various sorts) and applying these methods to three case studies of twentieth-century conflicts (namely, the US invasion of Grenada, the French invasion of Indochina, and the British invasion of the Falkland Islands).

The driving hypothesis of this study is that a combination of status-seeking behaviour and a system of subjective beliefs and identity traits reflective of what I call "nostalgic virility" together form an important set of influences that shape and factor into decision-makers' choices to engage their declining countries in wars within their periphery. The concept of nostalgic virility can be briefly defined as a system of beliefs and embodied identities referring to a particular understanding of a country's history and of notions such as power, bravery, strength, morality, and health.[10] Based on a rigorous and indeed surprising set of empirical findings, I argue that status-seeking and nostalgic virility are in fact more relevant to explaining these choices than the dominant explanatory frameworks of realism and the diversionary theory of war. Accordingly, before diving into a deeper explanation of the study and its findings, I will first briefly introduce the theoretical framework of status-seeking and nostalgic virility by situating it within the broader field of IR theory.

Situating This Study: How Have IR Theorists Sought to Explain War?

Realism

Perhaps the most dominant theoretical assumption used to explain the causes of war is realism (which I call here "traditional" or "mainstream" realism). Most realist studies (following Kenneth Waltz's influential 1959 book) are built on the argument that what matters for understanding international politics is the distribution of (military) capacities within the international system and the fact that the system is inherently anarchic, leading actors to trust only themselves for their survival.[11] This distribution, the argument goes, influences foreign policies generally, and decision-makers more precisely, in their choice of either diplomatic or coercive options.[12] One main issue with mainstream realist arguments (especially neo-realism), however, is that they are too mechanical: under this framework, a shift in capacities leads necessarily to a change in policy. As such, many scholars have developed competing theoretical frameworks structured primarily around questions of status.

Status and Identity

One recent critique levelled against realist frameworks is that they have not adequately integrated the role of identity and issues of status-recognition into their understandings – apart, perhaps, from the work of Andrew Greve and Jack Levy.[13] Despite Thomas Volgy et al.'s writings[14] about status differentiation and the role of status attribution in the production of different strategies, realist scholars remain silent about how to conceptualize status "identity" and recognition. They do not engage with the subjectivity nor with the affects and consequences produced by each status situation. Rather, they remain within the materialist/strategist understanding of status and do not discuss status in its dimensions of political subjectivity – i.e., from the ideational perspective of the state in question, and how this subjective dimension then alters the logics of status-producing strategies. Capturing this subjectivity would require adopting a different epistemological perspective that moves away from the premises of realism.

The importance of status, identity, and recognition as causes of foreign policies and wars has been addressed by scholars inspired by constructivism. The literature is particularly well developed on the impact status has on countries' foreign policies and on IR in general. For some, status is a strategic tool that provides a powerful deterrence (i.e., who would want to confront the military of a great power?).[15] For others, status impacts foreign policies because it is part of national identity, and as such, provides guidelines for action.[16] A related topic has also agitated scholars: is status achieved once sufficient military capacities are possessed, or is the process more complex, requiring recognition by relevant others? In the latter case, status recognition would require states to behave in certain ways and send appropriate status signals.[17] The ideas of recognition and of satisfaction/dissatisfaction of status attribution add new insights to our understanding of the causes of conflicts[18] by identifying various situations of status recognition denial.[19] Together, these studies point us away from the realist focus on the material conditions of power and towards an understanding of the importance of symbolic and subjective elements in the causes of international conflict and war.

Decision-Makers and Their Subjectivities

Another way of understanding and analyzing the causes of war is thus to place the emphasis of the study on decision-makers. The basic idea of this approach to conflict studies is to highlight the key importance of specific individuals within states' decision-making apparatuses.[20] In addition, it is also said that there is a tendency for people to identify countries with their leaders: if leaders are seen as tough, then their country is seen as such, including the inverse.[21] Among the most prominent of frameworks in this literature explaining why and how leaders matter is the model known as the diversionary theory of war (DTW). When leaders face a high probability of losing office, they can be tempted to initiate an international crisis. In so doing, they attempt to shift domestic public opinion away from domestic failures, betting that in a time of crisis people will put aside their differences, disagreements, and hostilities to support the leader through a "rally-around-the-flag" effect, thus creating a pressure and influence through an in-group/out-group dichotomy of support or loyalty.[22]

This model has been enriched by studies looking at how offices of power have been lost, finding that it matters whether removal from office happens due to a lost election, a lost vote of confidence, or a coup/revolution.[23] If leaders are at risk of being forcibly removed from power, they are more likely to initiate an international conflict to remain in power and thus avoid potential punishment. This has been referred to as a "gamble for survival."[24] Overall, then, the assumptions made in these models are that (1) decision-makers will take greater risks to *maintain* their domestic and international positions than to *enhance* these positions; (2) they find it difficult to accommodate political losses, often running high risks to recover them[25]; (3) leaders decide to go to war in order to avert or recoup losses, especially if national leaders have experienced a sharp decline in relative power or prestige; and (4) that seeking to recoup losses makes leaders more risk-tolerant and therefore more likely to engage in riskier policies.[26]

A second trend argues that what matters to the causes of war, when studied from the perspective of individuals, is who specifically the decision-makers are and what they think. Scholars working on the personal characteristics of decision-makers have tried to identify key elements to build an archetype of the war-prone leader.[27] Observers of international politics have also been interested in the desire for status and perceptions of virility, both of which, as we will see in this study, can impact decision-makers' decision to go to war.

These individual and subjective aspects have, however, been largely discarded throughout the development of traditional theoretical perspectives in realist thought. Some have gone as far as suggesting that values such as glory, honour, and courage are outdated as motivations to go to war.[28] Symbols, feelings, and perceptions have also been lost during the behavioural revolution in political sciences.[29] As a consequence, emotions have often been rejected as a legitimate object of research in IR, with many scholars holding that, methodologically, they cannot be studied rationally due to their "inherent" irrationality, except perhaps by studying them as limited and irrational exceptions to "normal behaviour"[30] or by treating them through strategic and rationalistic terms such as the pairings of fear/security and interest/power in deterrence theories.[31]

As such, emotions have been largely expelled from the study of IR and have mostly been present in cultural studies, especially of Arab and Asian politics.[32] Recent years have seen a resurgence of studies of emotions, however, following scholarly works in the fields of sociology and cultural studies.[33] Emotions are becoming more systematically and comprehensively examined to analyze politics and IR in general, and conflicts in particular, either by looking at the importance of emotions in suicide-bombers and terrorists,[34] looking at emotions felt by leaders (especially fear of humiliation, fear of appearing as emasculated, etc.),[35] or by looking at how emotions are integrated into foreign policies and expressed through diplomatic channels.[36]

The Argument in Brief: Status-Seeking and Nostalgic Virility as a Cause of War

This very quick overview of the literature shows that while the causes of war have been well theorized by scholars and observers of international politics, a burgeoning yet insightful set of more contemporary studies on the role of status and emotions is also being developed. These newer studies could benefit from a more systematic research methodology. Namely, scholarship on status differentiation and the role of status recognition in the production of different foreign policies has not really provided concrete and replicable research tools or conceptualizations for studying how the subjective dimension of status can alter the logics of states' (and, especially, individual decision-makers') status-producing strategies.

This study, therefore, combines the various traditions briefly outlined above while developing a comprehensive approach for understanding how status-related identity issues, combined with the personal belief systems that state leaders hold, can influence political decision-making on foreign affairs.[37] One main objective of this book is thus to contribute to current debates over the question of how actors and structures are co-constituted. Another is to examine whether symbolic views, conceptions of status, and perceptions of virility should be treated as important causal factors in the decision to go to war.

To this end, the main hypothesis of this study, tested in comparative perspective, is as follows: when faced with status decline and/or a challenge from a less powerful actor in their periphery,[38] leaders of great powers decide to react with a powerful military response due to a factor I term "nostalgic virility" – a belief system articulated around a nostalgic and virile understanding of what it means to be a great power. Nostalgic virility can be defined at the very least as a combination of the following traits: (i) an emphasis on physical strength; (ii) an emphasis on traditional virile values (e.g., firmness and bravery); (iii) the use of narratives that forward and enhance ideas of moral and behavioural superiority and set examples and models for others to follow; (iv) a positive understanding of a country's history and legacy, which serves as a reference point to assess and understand situations; and (v) a short time preference (i.e., any issue has to be addressed immediately).

As a subjective belief system, nostalgic virility acts as a filter according to which leaders assess various elements: what it means to be a great power; what their country's status is on the international stage; and what the appropriate actions/reactions are in given situations, such as how to react to a provocation made by a weaker actor. The nostalgic-virile understanding of great power status and its subjective filtering of decision-making criteria result, as I argue, in a situation wherein leaders, when faced with a challenge from a less powerful actor in their periphery, will react with a powerful military response that cannot otherwise be explained through the traditional theoretical approaches to IR and the causes of war. Nostalgic virility shapes a certain worldview and understanding of events that calls for strong-handed actions: the aggressor has not only to be punished, but punished severely, and not really because of the aggression in itself, but mostly because of its *symbolic aspects* – namely, its challenging of roles and duties that leaders of great powers have attributed to their country (or that they think others have attributed to their country), or its symbolism as a gesture of great power status recognition denial. The consequence of such challenges is that leaders (as well as their country) then appear as weak, uncertain, or "worse," effeminate or castrated and in decline. This image is accentuated by a glorified interpretation of the past: the fact that the country was not previously challenged by weaker actors in its periphery, but is now, implies that the country's symbolic deterrence through status has vanished and that the country has, in the eyes of others, become weak.

Status considerations and virile emotions thus interact. Through their subjective interpretations of what represents great power status, leaders tend to understand any aggression in their periphery by a less-powerful actor as an international insult (or even as a narcissistic insult). Such challenges demonstrate that deterrence by great power status did not work, and that their country (or even themselves as leader of the country) is no longer respected nor feared. It is not only the image that others have of their country that is at stake if no immediate and strong action is taken in response to the challenge, but also the image of the leader of a great power themselves that other leaders hold (and that the domestic opinion can also hold) that risks collapse.

In light of its theoretical positioning, this study contributes to the realist literature from a novel epistemological perspective by adopting an eclectic posture (inspired by Samuel Barkin,[39] and Richard Ned Lebow[40]) that is situated between constructivism and realism, putting the two approaches into conversation so as to offer a new take on the classical question of the causes of war. The eclectic posture of the study avoids limiting itself to the dichotomies between realists and constructivists by proposing a sort of interactionist view. Namely, by beginning with the premise that nostalgic virility is a potential *reason* for going to war, to which material reality must then align itself in order to lead to conflict, we may better understand how international aggression is provoked after leaders who are obsessed with nostalgic virility (such as the Iron Lady) are actually challenged, or when their power is actually "materially" in decline. In such cases, we find that their nostalgic-virile understanding of great power status causes them to view war as the only option for sending what Jervis calls "indices"[41] to the aggressor, and thus halting the decline process.

A Few Words on Research Design, Findings, and Relevance to Contemporary Conflicts

It is important to note that this research does *not* have the objective of proving that status-seeking behaviours associated to nostalgic virility are necessary conditions for deciding to go to war in general (i.e., in any case involving decision-makers of great powers against a less powerful actor). For this claim to be made, further research involving different and additional parameters would be required. Instead, the objectives of this research are: (1) to investigate whether the framework of nostalgic virility developed here can serve as a definitive explanatory reason for why decision-makers decide to go to war in their country's periphery against a less powerful actor; and (2) to demonstrate the relevance and benefit of this theoretical and methodological approach to the study of international relations generally, and of contemporary conflicts in particular.

This study employs a rigorous, systematic mixed-method approach to discourse analysis to investigate, using newly declassified documents, three case studies of great powers going to war in their periphery throughout the twentieth century. At the methodological level, primary sources have been under-used in past research, with most studies resorting to a "snowball" strategy of data collection to support their analyses, rather than basing them on a systematic review of primary sources (which are often classified and/or hard to come by). To ensure the validity of this study, I therefore use an extensive dataset of newly declassified primary sources (accompanied by secondary sources and academic/historical accounts for context) for my analysis, which allows for an unprecedented look at the discursive trends underlying the highest levels of decision-making during these conflicts (more details on these datasets and their construction can be found in chapter 2). The analysis of these datasets evaluates for evidence of three key models or theoretical frameworks: the realist-strategic framework, the DTW, and the novel framework of status-seeking and nostalgic virility. The evaluation of these three models is conducted without any a priori orientation, meaning that each framework is treated equally in the research design and analytical process. Results are not force-fit into pre-defined boxes; rather, the competing theoretical frameworks are put into conversation and tested equally across the three case studies. This is especially uncommon when it comes to studying emotions in IR, wherein researchers have tended to prefer single-n case studies and to investigate evidence for one single framework or another in isolation.

On the whole, the findings from my analysis reveal several interesting take-aways that indicate that, while realism and the DTW are at least partially capable of explaining the decisions to go to war, in each case the empirical evidence strongly suggests that it was a consensus among decision-makers (whether conscious or not) over deep-rooted concerns shaped by or representative of nostalgic-virile belief systems that acted as the prime motivators influencing the decision to go to war. The results therefore suggest that the combination of status-seeking behaviours and nostalgic virility offers the clearest and strongest explanation as to why these wars occurred, when compared to the existing common theoretical frameworks of realism and DTW.

However, nostalgic virility is, perhaps unsurprisingly, manifested in unique and nuanced ways within each case study. Although they share key similarities in the nostalgic-virile narratives at work in shaping the decisions to go to war, there are key differences showing that nostalgic virility is not monolithic. Each case offers internal variations in how the criteria of nostalgic virility are embodied by individuals. For example, in at least one case, nostalgic-virile elements were associated with strategic and geopolitical concerns typical of realist frameworks, while in another, nostalgic-virile elements were found associated with elements of the DTW. These synergies point towards opportunities for future research, especially in developing further precision with regard to how nostalgic virility can be conceived and expressed by individuals in a variety of complex ways. Accordingly, it is important to remember that even if status-seeking behaviours combined with nostalgic virility are identified here as the most important factors influencing decision-making in the three case studies, it is nonetheless just one factor among many within the messy ecosystem of influences, narratives, arguments, and belief systems shaping the decision to go to war.

We must note, therefore, that like any theory this framework has its limitations (discussed in more detail in chapter 2): it is possible only to argue that status-seeking behaviour combined with nostalgic virility was important in the three particular cases tested. In short, then, what must be taken away from this study is that emotions and a virile-nostalgic conception of great power status *can* play an important role in whether decision-makers go to war. Indeed, based on my findings it seems that these factors do play a very important role compared to others. However, whether emotions and a virile-nostalgic conception of great power status are the *most* important factors remains to be proven by further research and in different contexts, and – perhaps more interestingly – by applying this framework to more contemporary conflicts to determine whether it maintains relevance beyond the twentieth-century context.

For instance, as I discuss in more detail in the conclusion, the theoretical framework and methodology developed here have already been fruitfully applied to ongoing conflicts in the South China Sea and Ukraine, each of which have revealed new insights that can help policymakers, academics, journalists, and others to make sense of these complex international developments. Accordingly, the explanatory power of this framework is not, it would seem, limited to past wars initiated by great powers against lesser powers in their periphery.

In an era of increasing global turmoil, renewed frictions between China, Russia, and the West, and growing status conflicts between countries such as China, Russia, and Japan, among others, we are in desperate need of novel ways of understanding international relations as they unfold through both symbolic *and* material power struggles. It is hoped that this study will serve as one comprehensive attempt at such novel analyses that can spark further research and refinement of both theoretical and empirical models for studying status, identity, emotions, and subjectivity dynamics in IR. With this in mind, I will now briefly outline the structure of the book and how it explores the questions and hypotheses discussed above.

Outline of Book Structure

This introduction briefly introduced the reader to the overarching topic of this study, how it has been studied in the existing literature, and why a new research methodology and theoretical framework is required to advance our understandings. It also provided a concise overview of the theoretical traditions I will discuss throughout the case studies. The chapter then ended with a brief discussion of the main empirical contributions and broader relevance of the study.

In chapter 1, I outline in detail my theoretical framework that builds on questions of status and nostalgic virility. The reader will understand how status is conceived and operationalized in this research, how we can understand nostalgic virility as a potential causal mechanism, its impact on decision-making processes, and how nostalgic virility makes it very difficult for certain status-seeking individuals not to go to war when they are provoked by weaker actors.

In chapter 2, I outline the research design, methods, and instruments used to conduct this study. More specifically, I explain what data sources

I used, how and where I collected and managed them, and how I constructed my analytical framework through the use of a coding dictionary (i.e., a set of coding questions and protocols used in critical discourse analysis to ask a standardized set of questions equally for each document across a large dataset). Details about the coding process and the theoretical inspirations behind the coding questions are also presented in greater detail in the methodological appendix.

Chapters 3–8 comprise the core of this study, outlining the findings for each of the three case studies in turn. Namely, chapters 3 and 4 cover the French invasion of Indochina, chapters 5 and 6 cover the British invasion of the Falkland Islands, and chapters 7 and 8 cover the US invasion of Grenada. Each pair of chapters follows a similar structure, which I will avoid repeating three times here. In particular, the first chapter for each case study (i.e., chapters 3, 5, and 7) introduces a brief historical overview of the conflict under analysis, followed by outlining some a priori evidence in favour of each theoretical framework, as found in the existing literature. The remainder of these chapters then discusses the findings for the realist-strategic and DTW frameworks. The second chapter for each case study (i.e., chapters 4, 6, and 8) then outlines and discusses the findings for the status-signalling and nostalgic virility framework, respectively, followed by concluding thoughts that place these findings in comparative perspective with the findings from the other frameworks.

Finally, chapter 9 concludes the book by offering a comparison between the results obtained in each case study and identifying key similarities and differences. Then, the chapter builds on this to argue how these results help us to better understand great powers' behaviours, as well as international relations more broadly. The chapter finishes with a discussion of how this theoretical and analytical approach can be applied to contemporary conflicts by summarizing the results of two recent studies using the same coding dictionary – one on Russian discourses about Ukraine since 2012 and another on Chinese discourses about the South China Sea since 2014 – and the implications of the nostalgic virility framework for better understanding these ongoing developments.

1

Status-Seeking and Nostalgic Virility

Towards a Novel Theoretical Framework for Understanding the Causes of War

As a basis of this study and its contribution to the field of IR, I am examining the co-presence of both *symbolic factors* (e.g., status and perceptions of status) and the *logics* of what I term "nostalgic virility." Nostalgic virility can be defined, at the very least, as a belief system that combines the following traits: (i) an emphasis on physical strength; (ii) an emphasis on traditional virile values (firmness and bravery); (iii) the use of narrative that emphasizes moral and behavioural superiority and sets examples and models for others to follow; (iv) a positive understanding of a country's history and legacy, which serves as a reference point to assess and understand situations; and (v) a short time preference (any issue has to be addressed immediately). This belief system acts as a filter according to which leaders assess various elements – for example, what it means to be a great power; what their country's status is on the international stage; and what actions are appropriate in given situations, such as how to react to a provocation made by a weaker actor.

This model is in fact built around a synergy. I argue that the nostalgic-virile traits develop an explosivity once combined with traits of power, namely with great power status. So, what are these traits and where do they come from? I will first begin by revisiting the notion of great power status and its implications before moving on to outline in greater detail the concept of nostalgic virility and how it can maintain interactive synergies with status perceptions and, ultimately, foreign policy decision-making.

Defining Great Power Status: Symbols and Expectations

Great Power Status and Its Recognition Process

"Great power" is a concept used by IR scholars to single out or refer to the most powerful states around the globe. Although it is a broadly used concept, there is no definitive consensus on its definition. A popular definition follows the realist understanding of great power, which puts the emphasis on having important economic and military capacities, as well as a large population and territory.[1] These elements come together to describe great powers as states that can win major wars,[2] that can rely on their own capabilities to provide for their own security,[3] that should only fear other great powers or alliances of other powers,[4] and that seek to influence the course of international affairs relatively independently of other major powers.[5] This view of great power has faced much criticism because it leads to an overestimation of the importance of military capacity,[6] despite the fact that history provides many examples of the importance of *power perceptions* (as opposed to material capabilities alone).[7]

The definition of great power status must incorporate *both* the importance (and use) of material capacities *and* the importance of perceptions of such capacities. To define great power status, this study therefore builds on the three definitions of power given by Steven Lukes,[8] by Michael Barnett and Raymond Duvall,[9] and especially by Manjeet Pardesi.[10] The main elements of these definitions include: (1) power as interactions of *direct control* to make others take (or prevent them from taking) decisions; (2) power as *influencing* the decisions of others or inciting them to take decisions (e.g., by influencing their perceptions or values); and (3) power as *creating social norms.*

Based on these elements, a great power is, at the very least, a country possessing the following elements:

(1) *Having a material presence in regions other than its home region.* This is materialized by economic and/or strategic (military) interests. It grants the country visibility on the international stage as well as influence over a region or another country. It can be materialized by the number of diplomatic representations and/or diplomatic missions (embassies, consulates, cultural centres, archaeological

missions, etc.), military bases,[11] or specific economic and political agreements giving the state an influence of some sort within the domestic affairs of another.[12]

(2) *Power projection capacities.* This derives from the previous point. Having interests in other regions supposes the capacity to protect and even to expand through overseas military bases or aircraft carrier strike groups, for example. The essential idea is that the country has capacity to intervene in other regions. A great power is autonomous and enjoys freedom of action: it defines foreign policies without having to ask permission from another great power or mirroring what others do.

(3) *The will and capacity to manage the international system and enforce its rules.* Great power status is a particular type of status: it goes hand-in-hand with expected higher-task performances and a better ability to solve regional and international issues.[13] Since the Vienna Congress (1814–15), great powers have granted themselves this attribution and the international community has recognized the existence of their specific responsibilities, which change with the global context: preventing further wars from breaking out and tackling future international crises[14]; and preventing or punishing physical aggression,[15] contraventions of international law,[16] and violations of human rights – as embodied in the recently popularized concept of a responsibility to protect (R2P).[17]

(4) Fulfilling these three criteria brings about a fourth: *status recognition.* The literature emphasizes three key audiences to convince in order to obtain and retain status recognition. The first is status self-attribution by either decision-makers and/or domestic opinion.[18] Here, the insistence that their country should be seen as a great power and granted privileges and prestige linked to that position becomes a key element of a state's foreign policy until such time that full recognition occurs.[19] The second important audience is peers (i.e., other great powers), in what is termed peer co-optation. This usually depends on the capacity of the claimant to send appropriate status signals to their peers,[20] for example by performing roles or satisfying other criteria associated to that status.[21] The last important audience is the international community more broadly, which usually follows from the status

recognition given by other great powers (i.e., peers), as the international community tends to mirror what they do.[22]

The Importance of Symbols to Great Power Status and the Case of the Periphery

Once status is achieved, it is essential to maintain it – and this demands constant effort. Status attribution is highly subjective and depends on how others see a given situation. The perception of fulfilling status criteria is done through reference points by which a situation or an actor's behaviour is analyzed. Troubles arise when those reference points used by the status claimant differ from those used by others.[23]

If a state's behaviour appears ambiguous compared to its claimed status, it might encourage others to stop perceiving it as a great power, or consider it to be in decline. When a direct challenge or provocation occurs, what is at stake, symbolically, is the ability to meet certain status criteria. If the challenge does not receive the correct response (based on great power status expectations), it could increase status uncertainty and ultimately lead to status recognition denial.[24] If the challenge is addressed correctly – in sync with status expectations – then there will be no status uncertainty (or if there was any before, it will probably stop). Different actions can be employed to put a stop to both status uncertainty and status denial, but it "often requires publicly enduring some cost or risk, often by participating in violence,"[25] such as military intervention in the periphery.

The meaning of "periphery" is both ambiguous and polysemous. In its ancient Greek sense, it meant what is on the outside, what surrounds a centre, or the external boundary or surface of a body: the outward bounds of something as distinguished from its interior regions or centre; an area lying beyond the strict limits of a thing. Periphery is often defined by opposition to a centre that is more developed and wherein everything core or important happens.[26] In IR, the word periphery was first used by Marxist scholars as a synonym for Third World,[27] before being transposed to the field of asymmetric conflicts.[28] In strategic terms, periphery is understood as being composed of "areas of the world considered to be of secondary or tertiary interest."[29] It is opposed to core interests considered vital and essential that must be preserved at all costs in order to sustain the survival of a state and its security.[30]

I have chosen to build on these early definitions of periphery but understand that it is not a fixed term. What belongs to the periphery might well become part of the core and vice versa: "An area that is believed to be of inordinate importance at one moment may well become less so in later years."[31] It depends on policymakers' definitions of a state's interests and strategies. To summarize, then, periphery can be defined according to the following four elements:

(1) *Geographic distance from the centre and/or vital interests.* The idea of pure geographic distance is too restrictive and needs to have a broader understanding. To illustrate, take Singapore and the Ionian Protectorate: both are located far from the centre (the UK), but – despite being the closer of the two – the Ionian Protectorate was less important to Britain than was Singapore. Singapore might be a long way from the UK geographically speaking, but it is positioned close to a location of core interest to Britain (India) and controls the strategic region of the Malacca Straits.

(2) *Political distance.* This point is a consequence of the previous one: the periphery is politically distant from the centre, in the sense that little attention is given to what happens there and it is often marginalized in political discourses (e.g., it has an absence from political debates and strategic planning, the government is not keen to commit more resources, or domestic opinion shows little interest).

(3) *Small level of strategic threat.* Following Taliaferro's comments,[32] any challenge or threat in a great power's periphery cannot harm nor threaten its core interests or its existence as an independent state. If this were the case, the challenge would not be simply peripheral but rather central to great powers, who focus first on their vital interests and peer competitors.[33]

(4) *Symbolic/emotional connection to a great power.* Peripheries often have a symbolic meaning for a great power because they have been part of its empire or sphere of influence. If we look back through history, peripheries were used to acquire prestige in preparation for struggles in the centre,[34] just as in the cases of Julius Caesar and Napoleon Bonaparte. Military interventions in the periphery also have a symbolic connection to the status of great power itself: a proof of power projection capability and of being committed to the special duties of great power status.

On initial examination, a challenge in the periphery is not a cause for great concern,[35] especially if the challenger is a less powerful actor. A priori, the power gap between the two leaves little doubt about how the challenge will end: the foolish challenger will be put back in its place by the great power, in theory at least. From a symbolic point of view, however, the challenge might be more important than it seems. Intervening in its periphery is one way for an actor claiming great power status to signal their capacity and resolve – i.e., their continued embodiment or merit of their great power status. As such, status considerations are key contenders in the explanation and understanding of why great powers choose to go to war in their peripheries.

Towards a Nostalgic-Virile Understanding of Great Power Status: No One Wrongs Us with Impunity

IR scholars have agreed that there are three main options when reacting to any given situation: (1) to not react at all and ignore the situation; (2) to react directly (e.g., going to war); or (3) to react indirectly. However, given the inherent complexity of international relations, scholarship has remained contested and unclear in regard to *what* would make a decision-maker decide on a particular option, or why they would decide to react immediately or at a later time.

A military reaction to a challenge in the periphery is likely to occur if leaders have a nostalgic-virile understanding of great power status. Virility makes the idea of being challenged by a less powerful actor unbearable, while nostalgia influences how leaders choose to respond to a challenge more generally. Building on the discussion of great power status and its role in the decision to go to war, the remainder of this chapter will outline the ways in which nostalgic-virile individuals (namely decision-makers) influence and are influenced by conceptions of great power status and, in turn, how this combination of perceptions and identities or subjectivities can impact a country's decision to go to war in its periphery.

Defining Virility: From Virile Individuals to Virile States

The idea of virility is strongly linked to the male. The Latin word for virility stands for both the age of a man and for masculine organs,[36] while the expression *virus vir*, for example, means real man/husband.

In Rome, and then during the Middle Ages, *virtus* meant strength and bravery in war, but carried with it other traits such as prudence, justice, self-control, and courage – all characteristics a reputable Roman man was expected to have.[37]

Virility is associated not only with the idea of muscular strength, but also with a strong connection to health. This connection has been around for centuries: "In Rome, virility is characterised first by an active and/or passive sexuality [active meaning penetrating and passive meaning penetrated], all the more so because some people contest this idea of passive attitude."[38] The word *virtus* also conveyed specific traits to which every Roman male should have aspired: prudence, justice, self-control, and courage. Historians have reported how Octavian used these traits to his advantage to denigrate Mark Antony, the latter reportedly bewitched by Cleopatra VII.[39] In Octavian's propaganda, Mark Antony was no longer a true Roman either physically or morally, and therefore had to be vanquished.

In ancient societies such as Egypt, male health, virility, and physical prowess were also often examined through rituals. The Sed Festival, which originally celebrated the potency and continuing longevity of a Pharaoh's rule, was transformed into a rite of royal regeneration.[40] It was observed after the first thirty years of rule and then every three years thereafter, symbolizing to the people that their ruler was still capable of ruling the kingdom. This was especially important for an aging Pharaoh (e.g., Ramses II who lived to seventy-four) or Queen (e.g., Hatshepsut). Tests of physical ability were often held, such as running, hunting lions or hippopotami, or capturing a bull. During proceedings, the aging Pharaoh was symbolically killed and buried and his personal connection to the gods re-established.[41] From then on, a ruler's physical abilities were scrupulously observed and commented upon.

After millennia of changes and developments to these virile understandings of power and leadership across Western cultures, the narratives, rhetoric, and symbolism associated with a leader's body arguably reached their height of sophistication under the French Monarchy.[42] During the eighteenth and nineteenth centuries, moral strength was regarded as particularly important: men's ethics were based largely on the idea of respect and appropriate behaviour.[43] Any challenge to one's honour demanded compensation, but resorting to the law to seek redress was considered a bad idea – especially in a society that would have regarded such action as a public acknowledgement of two damaging facts: (1) confessing publicly to having been offended, and (2) "avoiding one's manly responsibility and compromising even more seriously one's honour by agreeing to sell it for a sum of money."[44] In the eighteenth and nineteenth centuries, the way to solve the matter was by duel: a "game" of honour where the outcome did not really matter. To refuse a duel was the worst of all possible scenarios as it "is to open oneself to the suspicion of cowardice and risk, for that reason, being denied the status of a man ... To be no longer a man is to risk social death."[45] If someone refused to defend his honour and status, the thinking went, how could he possibly defend those placed under his responsibility (e.g., his wife, children, or even country)?

Although less prevalent than in the nineteenth century, some countries still regard the physical abilities of their leaders as important. Vladimir Putin is a prime example.[46] His holiday photos are particularly revealing of his efforts to show the Russians that he is fit to lead: namely, that he is strong, brave, and daring.[47] In another instance, the Kremlin widely publicized the fact that he had received a nude calendar from some female students of Moscow University, thus "proving" that he continued to be sexually attractive in spite of his age.[48]

Most understandings of virility emphasize the traits mentioned above: being strong, protective, brave, and competitive.[49] These traits tend to be associated with successful leaders, or at least are often expected of them.[50] To show strength and virility – and that they "have balls" (i.e., are tough, daring, and decisive) – male leaders draw attention to their bodies to prove or demonstrate potency, often using somewhat crude banter in doing so.[51] But what about women?[52] History has many examples of what could be termed "virile" or "phallic" women leaders.[53] When in power, women can have as much ambition, hunger for power and control, and fearlessness as men. What may differ is the way men and women choose to *display* these characteristics.

Women might stress their toughness through attributes traditionally associated with femininity, such as care, empathy, and maternalism.[54] They may project their femininity and maternity as an image of national renewal. One of the best examples is British Prime Minister Margaret Thatcher, who liked to remind people that she passed the bar exam all while taking care of her young children. In doing so, she symbolically showed that accomplishing difficult tasks was not a problem for her, that she was not going to make any concessions, and was therefore capable of being tough.[55]

With conceptions of virility being historically very focused on the personality and traits of the leader, we may ask whether the idea of virility can be transposed from individuals onto states? How can the criteria of virility be understood from a state-centred perspective? If we make a basic comparison between individual and state – akin, perhaps, to a Hobbesian understanding of the body politic – the equivalent of the head would be the decision-makers, the equivalent of muscles would be the armed forces, or more generally the apparatus of coercion, and the whole of the body would be the population.

A virile leader is expected to take difficult decisions and successfully defend national interests to make sure that their country suffers no humiliation and will not be weakened on the international stage.[56] To ensure this, they might be expected to punish any actor responsible for causing national humiliation. In this context, the leaders' actions are highly important and symbolic: when a state's leader is considered strong (or weak), their country is too. In the 1980s, the question of a leader's health was particularly pertinent in the USSR when there was a succession of physically weak, sick, and older general secretaries at the helm: Leonid Brezhnev, Yuri Andropov, and Konstantin Chernenko. Time after time, they were all questioned about their ability to lead and make difficult decisions.[57]

The belief that a country is weak or strong is enhanced by its "performance" on the international stage: it is weak if defeated in war or diplomatic crises, strong if it is victorious. Cynthia Weber even compares the succession of US "defeats" in the Caribbean to castration: "America's body politic has been decapitated. Headlessness marks its Caribbean castration ... Left headless and thus voiceless, the United States no longer can be assured the last word in global affairs. Its self-appointed role as keeper and enforcer of the law of the father is now in jeopardy."[58] US "castration" went further still under the presidencies of Nixon and Carter, who saw their attempts to "rephallusize the US" fail. Under Nixon, US muscle power (the armed forces) appeared helpless as communism was spreading in South America (Salvador, Nicaragua), while the US "head" was also discredited during Watergate, before being further humiliated in Iran.

In some cases, leaders have conflated the people's virility with that of the country. Birth rates, sports, contraceptive practices, and authorizing and legalizing sexual behaviours judged as "deviant" (e.g., homosexuality) are all relevant here. For example, French governments attributed the defeat against Prussia in 1870 in part to a lack of physical strength among French soldiers. To remedy the situation, physical education classes were established as part of the public curriculum.[59]

The biopolitical idea of strengthening a country's virility and making a "new man" emerges as central to totalitarian regimes.[60] A horrific example is the Nazi attempt to "regenerate" the German population from 1933 to 1945. On 15 September 1935, the infamous *Gesetz zum Schutze des deutschen Blutes und der deutschen Ehre* (Law for the Protection of German Blood and German Honour) was passed, forbidding sexual relationships between Aryans and people judged to be "inferior." Evidence from Nazi policies and discourse also reflects their desire to eradicate any type of "weakness" (e.g., handicapped or deviant individuals): "If one million children are born every year in Germany and if one eliminates the seven to eight hundred thousand weakest, a potential result would be that the country's strength would grow."[61] The Nazis eventually opened "reproduction" centres – the *Lebensborn* – with the aim of ensuring the birth of "superior" individuals destined for the SS.

Looking back through history, we see that there is a long tradition of virile thinking and of incorporating virility into social and political behaviours, a history that started in antiquity and developed and morphed into particular socio-historical forms well into the nineteenth century, ultimately leading to an apex in twentieth-century fascist and totalitarian regimes such as Nazi Germany. Given the fact that conceptions of virility are not only specific understandings of physical abilities, but also of behaviours considered as canon and appropriate, virility has longstanding implications in the history of IR.

Defining Nostalgic Virility: Strength, Punishing Deviant Behaviour, and the Past as Guidelines for Action

My definition of nostalgic virility and its criteria draws heavily on the elements described above and attempts to operationalize them. Nostalgic virility can be defined, at the very least, as a combination of the following five traits: (1) an emphasis on physical strength; (2) an emphasis on traditional virile values (e.g., firmness and bravery); (3) the use of narratives that enhance moral and behavioural superiority and set examples and models for others to follow; (4) a positive understanding of a country's history and legacy, which serves as a reference point to assess and understand situations; and (5) that nostalgic virility implies a short time preference (i.e., any issue has to be addressed immediately). These elements can be expressed differently and are found in different forms and configurations depending on many factors, such as the individual's political culture, context, and history. However, the diversity of real-world manifestations of nostalgic virility can, overall, be reduced to these five general elements.

For many nostalgic-virile leaders, more intense virile values based on the idea of punishing deviant behaviours and a specific relationship toward the past can be identified behind the simple façade of displaying strength. Not all virile individuals endorse the more intense virile values, nor are they all nostalgic, but the combination does often occur. Indeed, many logical ties unite these elements: the need to appear strong and resolute is assessed through comparisons with key moments in a country's history (either to repeat or avoid them), and the necessity of punishing morally deviant behaviour can be seen as the most effective way (if not the only one) to restore the past or fill the gap between a nostalgic and great past and the desired present. These elements can present themselves positively or negatively. Positive displays of strength, for example, place an emphasis on being strong (physically or symbolically), while negative displays of strength emphasize a rejection of weakness – refusing to "lose face," be humiliated, or appear feminine.

Let's explore the five core elements of nostalgic virility in a little more detail.

(1) *Emphasis on physical strength.* This can be shown in many ways. Individuals can emphasize their physical abilities by displays of intensive physical exercise or by drawing attention to their muscular physique. Leaders can also display physical strength at a symbolic level: visiting a war zone or appearing in military uniform to reinforce their position as commander-in-chief, as did President Bush when he went aboard the aircraft carrier USS *Abraham Lincoln* in 2003. They can also draw attention to the state's "muscle," namely

the military, by praising its cutting-edge equipment, training, and successes, for example.

(2) *Emphasis on "traditional" virile values.* Leaders want to appear in command: they know what to do and how to do it. This supposes two things: (i) *Being firm* – i.e., not making (or rarely making) U-turns, sticking to decisions no matter what, and interpreting that a change of mind might appear weak and irresolute. For example, during the 1981 Irish Hunger strike of IRA members imprisoned in Belfast, Margaret Thatcher refused to allow any concessions to be made. The strike eventually ended after the death of ten prisoners and when the remaining prisoners ended their strike. (ii) *Being brave.* – i.e., taking difficult and unpopular decisions and highlighting how difficult, albeit necessary, the decisions were.

In addition, I would argue that virile individuals have what Stephen Benedict Dyson calls a "low level of conceptual complexity," defined as the "degree of differentiation an individual perceives in the environment in which he operates."[62] It does not mean that individuals are unable to understand complex situations but rather that they tend to interpret them according to a "black-and-white" or binary information-processing style. Actions are seen as positive or negative, right or wrong, moral or immoral: there is no middle ground. Low complexity individuals see themselves as operating in a zero-sum game, one in which their gain or loss is exactly balanced by the gain or loss of others. I would add to Dyson's argument that, in some cases, low-level conceptual complexity can go as far as having a strong belief in natural laws: individuals believe that there are immutable laws that govern human affairs, such as "the strong do not get attacked but the weak do," "democracy means freedom and liberty," and "dictatorship means war."

(3) *Enhancing moral and behavioural superiority.* In other words, proving "fit to lead." This argument can also be made at a symbolic level, where the idea of health is linked to a political and/or moral feeling of superiority. The country is "fit to lead" because of its advanced political regime (in the case of a democracy) and/or because of its scrupulous adherence to legal rules and procedures, either at the domestic or the international level. Examples of being fit to lead might be when a country displays appropriate and civilized behaviour (e.g., negotiating instead of going to war). When and

if war becomes inevitable, "fit to lead" means that the rules of engagement given to the military are in accordance with international law (including attempts to confine military operations, avoiding an expansion of the conflict, trying to keep casualties to a minimum, especially among civilians, and not using nuclear weapons).

The criteria of being "fit to lead" can also be more diffuse and can constitute the root of interventionist policies, especially when these policies operate biopolitically or incorporate a biopolitical dimension. Such policies articulate governmental rationalities (and sometimes narratives) with the stated aim of providing for the health and welfare of populations.[63] Visions of enhancing moral and behavioural superiority, with or without biopolitical rationalities and narratives, articulate well with systems of metaphors identified by George Lakoff, who argues that our political worldviews are structured around implicit metaphors.[64] His study of these metaphors is centred on the context of American domestic politics and was organized around two main ideal-types: the "strict father" and the "nurturant parent." The "strict father" ideal-type essentially characterizes US conservative mindsets, while the "nurturant parent" ideal-type is found mostly in liberal mindsets.

The strict father mindset is built around a punitive sensitivity: individuals need to follow strict rules, cultivate self-discipline, and maintain a profound respect for authority. There is a strong moral dimension associated to this, as "deviant" behaviours are viewed as unacceptable and must be severely punished. In contrast, the nurturant parent mindset emphasizes the need to show empathy and to support and protect others. In the context of international relations and foreign affairs decision-making, we can understand these ideal-type mindsets and the metaphors that inform them as follows: a "strict father," punitive-type virility places the emphasis on the punishment of morally and behaviourally deviant activities such as committing genocide, breaking international norms, and/or acting with cowardice; by contrast, a "nurturant parent" or nurturant-type virility puts the emphasis on helping others to grow and become virile through a positive power relationship – what matters most is creating a safe and secure context for development.

(4) *Nostalgia*. History is a powerful source for providing guidelines for action. Scholars interested in public policy and crisis management have well identified the impact of previous experiences and analogies in the decision-making process,[65] as well as in the formation of identity.[66]

I argue that nostalgia and virility are traits that often go together. Nostalgia presupposes that leaders value their country's past and its legacy. It implies that leaders view the past in a positive way, especially their country's previous achievements. "The good old days" are remembered with pride. The past is viewed as something to be built on; it provides a guideline on how to be virile and act on the international stage. This is particularly the case if individuals perceive that something has recently changed, if they feel their country is not what it once was, or if their country is less respected and/or less efficient on the international stage.

Nostalgia tends to make individuals compare their country's roles and status with what it was in the past, especially during a "glorious" time. Nostalgic leaders, therefore, use historical analogies to point out what makes their country special – victory in war, for example – as a reference point to analyze and compare their current situation and status position. On the other hand, if a country is in decline, then the past can provide a toolkit to mirror and repeat previous actions and thus return to glory. Military capacity is often compared to what it was previously, and there are comparisons especially between what the military was able to do in the past (e.g., distant power projection) and what it is able to do now.[67] Similarly, if their country has been granted great power status in the past, virile nostalgic leaders would not be likely to accept any downward shift in their current status, and even less likely to accept any uncertainty about their legitimacy to be a great power. As John Darwin notes with regard to Britain, the glorious past is a powerful motive for refusing any status shift, even after a humiliating defeat: "Settling instead for a secondary role in the world was not easy to square with the service and sacrifice of Dunkirk, the Battle of Britain, the Blitz, Alamein or the D-Day invasion."[68]

(5) *A short time preference*. This focuses on the window of time in which leaders must act and take decisions. Virile nostalgic leaders

are likely to have short time preferences when challenged, as they want to respond immediately. Wasting time could prove potentially dangerous and delayed action interpreted as weakness (similar to refusing a dual in the eighteenth and nineteenth centuries).

On the basis of these five key indicators or criteria, it is important to examine the mechanisms that lead to nostalgic-virile individuals having the desire, and even the need, to punish or protect. I will examine these mechanisms in greater detail in the next section and throughout the case studies.

Nostalgic Virility and Challenges in the Periphery: Towards an Explanatory Framework

A challenge in the periphery is more than just a challenge presented to a great power, namely because of the connection that the periphery has to great power status. Periphery in itself symbolizes each of the criteria required to be a great power. The symbolic significance of a challenge in the periphery is increased if the challenger is much less powerful. Nostalgic-virile leaders tend to consider it as a deliberate insult to their country's pride and standing, as well as a denial of its superiority. The attack by a weaker actor is in itself insulting and humiliating and, under normal circumstances, great power status should have prevented it due to power disparity.[69] Humiliation grows in proportion to the roles attributed to a great power and what represents the opponent: it is hard for a democracy to accept being challenged and defeated by a dictatorship, especially if the latter has a terrible human rights record.

The feeling of humiliation is accentuated by various aspects of the virile understanding of great power status. Individuals perceived as physically strong, virile, and powerful do not usually come under attack. Strong and powerful states are not often challenged by weaker actors, and, in this virile view, those that have been but are no longer genuine great powers have become weak. Being challenged by a weaker actor is thus a symbolic contestation of a state's status, role, power, and – more generally – its virility. Nostalgic virility links challenges with clear-cut situations based on a black-and-white interpretation. The specific case of the challenge will be incorporated into a bigger, more global and

Table 1.1 Indicators of nostalgic virility

		Operationalization	
Elements	*Traits*	*Actions and behaviour*	*Examples of discourse*
Emphasis on physical strength	(i) Appearing personally strong and/or healthy	(i) Enhancement of body/physical ability (e.g., Putin) (i) Drawing attention to paternity/maternity (e.g., Thatcher)	(i) Pro-family/against homosexuality (in some cases, "femiphobia")[1]
	(ii) Appearing as symbolically physically strong	(ii) Appearing as commander-in-chief	(ii) Praising the military/praising power of the country/saying how powerful the army is
	(iii) Refusal to appear weak/effeminate/humiliated	(iii) Strong reaction to humiliating situations if no honorable exit (either for themselves or their country)	(iii) Claiming policy must be seen through because of promises made (iii) "This was a promise made to us and was not kept, therefore we react'" (iii) "If we do nothing, we will lose honor"/ "This is humiliating our country, we have to react" (iii) Punitive tone[2]
Emphasis on moral and behavioural superiority	(i) Appearing healthy/fit to lead at the personal level	(i) Enhancement of perfect health (e.g., Putin, opposing models Yeltsin, Andropov)	(i) Anti disease (e.g., fight against alcoholism/drugs)

/continued

1 Ducat, *The Wimp Factor*, 24.

2 Lakoff, *The Political Mind: Why You Can't Understand 21-st Century*; *Moral Politics*; *The Political Mind: A Cognitive Scientist's Guide*.

Table 1.1/*continued*

Elements	Traits	Operationalization: Actions and behaviour	Operationalization: Examples of discourse
Emphasis on moral and behavioural superiority /*continued*	(ii) Appearing healthy/fit to lead at the symbolic level of the state	(ii) Emphasis on the moral/political superiority of a country to lead the world. (ii) Scrupulous adherence to domestic and/or international rules. (ii) Sanctioning of perceived deviant behaviors based on an understanding of the criteria above. (ii) Developing policies having a biopolitical aspect and/or objective[3]	(ii) Arguments of superior political institutions. (ii) Arguments emphasizing that there are rules which must be followed (e.g., "We will limit ourselves to," "We will follow the UN resolutions," "We are committed to find a peaceful settlement"). (ii) Arguments emphasizing the need to restore conditions required for humanitarian assistance and economic development; (re)building a "proper" politico-judicial system in line with Western standards.
Emphasis on traditional virile values	(i) Firmness	(i) No U-turn on policies/standing firm even if facing contestation (at domestic or international level)	(i) "We shall not retreat," "This is a test of our resolve"[4]
	(ii) Bravery	(i) (ii) Use of force to protect (perceived or real) interests	(i) (ii) National interests mentioned, liberal internationalist discourses as justifications for military interventions

3 De Larrinaga and Doucet, "Governmentality, Sovereign Power and Intervention."
4 Tang, "Reputation, Cult of Reputation, and International Conflict," 43.

Elements	Traits	*Operationalization* Actions and behaviour	Examples of discourse
Emphasis on traditional virile values/*continued*		(ii) Taking difficult/risky decisions (even if opinion seems to be opposed to it regarding opinion polls)	(ii) Telling how hard the task was/telling how nobody had dared do something similar before
	(iii) Low level of conceptual complexity	(iii) Low complexity of individual's information-processing style	(iii) Them vs us; good vs evil; victory vs defeat
Nostalgia	(i) Positive understanding of history and of historical legacy	(i) Search for symbols of power and symbols of splendor (e.g., military build-up, costly armament programmed even if not relevant to the country's strategies	(i) Historical analogies, comparison between past and current situation/positions (i) References about glorious past of the country/stressing how the country has special qualities that set it apart from others (i) Mentioning that the current decline will stop and the country regain its status/power, etc.
Time preference	(i) Short time	(i) Use of expeditious means to solve issues	(i) Cannot stand humiliation by an inferior actor and seeks immediate action

dichotomous conflict: democracy versus dictatorship, rule of law versus authoritarianism, civilization versus barbarianism, etc. The failure of a great power is not only the failure of a specific country, but also of other countries with which it is associated.

When a challenged country is in full possession of its capacities, it is no real surprise to see it intervene. But what about great powers suffering decline (real or perceived, partial or complete)? Great powers are expected to win, and to win easily because of power disparity – just like every Spartan was expected to easily kill a zealot during the "Krypteia." Decline changes everything, especially because decline is often answered by reductions in military capacities.[70] Decline makes the military response and punishment more delicate and riskier. Failing to punish the challenger, or worse, defeat in war – like being caught cheating or failing during the "Agoge" and "Krypteia" – leads to denial of citizenship as well as public shaming and humiliation.[71] This begs the question: *If we don't react, what will the others think of us?* "Us" refers both to the leader(s) and to the country in general: *What would public opinion be about us as leader(s) and what would the other international actors think about us as a country?*

Nostalgic-virile leaders fear one thing in particular: to be seen or perceived as weak or, even "worse," effeminate or castrated. This could have future repercussions for them (losing power) and/or for their country (being challenged again); there could also be a cost to pay on their great power status (status uncertainty/status denial), and likewise, backing down could create a precedent.[72] The costs of backing down are, therefore, simply too high. Even if risky, nostalgic-virile leaders will gamble with their great power status and opt for a military response. They would risk national humiliation and the potential loss of great power status in order to end current status uncertainty/denial, and to show their country that it is not merely playing at being a great power – it *is* a great power.

Going to war in the periphery could also occur in case of a shift in great power status attribution or to a great power's criteria (e.g., being a liberal democracy). Nostalgic-virile leaders are less likely to accept such a shift if it put their status legitimacy claims at risk. In this context, they could interpret the new criteria as an implicit denial of status legitimacy, and decide to go to war to reaffirm their country's possession of great power status criteria. To do so, they might decide to intervene militarily in an already existing and long-lasting crisis/conflict (especially if it was in deadlock before their intervention), or they could try and thwart the foreign policies of great powers that have initiated and/or supported the shift in status criteria. Virility also has a relatively similar impact as religious narratives in the legitimization of a resort to physical violence: "[it] accentuates the non-negotiable dimensions of a conflict and its perception as a threat to the core of an individual's and/or group identity."[73] For the same reasons, once a crisis has broken out, they would be less likely to accept a compromise or engage in negotiations.

The first objective of a military response would be to reassure domestic opinion that the country still has the means of its claimed status, that it continues to be powerful, and that it is managed by competent leaders. Going to war also enables the decision-makers to prove themselves virile. They almost certainly will not actually fight on the battlefield, but going to war means they can appear able to make courageous and tough decisions, to show resolve and firmness (in not making any U-turns), and to engage their country in a war, something which is never easy.[74] Leaders demonstrate their potency, too, by becoming commander-in-chief and setting up the overall objectives or by arbitrating between the various operational options the military presents. It shows firmness in punishing the aggressor not just for what they did, but for the symbolic meaning of the challenge. The second objective would be to impress, or at least (re)legitimize, great power status and the roles associated with it in the eyes of peers (other great powers) and the international community. Warfare demonstrates in a concrete way that the country's military is competent, well-trained, well-equipped, and well-commanded. It can cause a reassessment of the superiority of the country, especially if the task force being sent is impressive in terms of number of units deployed (even if this is sometimes disproportionate in order to secure a quick and expedient victory against a hapless opponent).

There is only one way of avoiding (or ending) status uncertainty: by sending appropriate status signals. The question remains, however, as to what form the signals should take. If virility provides the required action, nostalgia provides the means. To get great power status recognized again, nostalgic leaders will choose a way to respond to the challenges that brings back imaginaries of the "good old days," that is to say the days when the country was fully recognized and uncontested as a great power.

Nostalgic-virile leaders look positively to the past and consider it as something to be built upon. Magnification of the past has a direct impact on present and future realities. Retrospection on achievements or glorious national successes serve as a reference for great power status legitimacy claims. Based on previous achievements – victory in wars, international posture, etc. – nostalgic-virile leaders believe their country deserves great power status. As John Darwin notes, there was a "belief that Britain holds a special role in the world … This was only partly a residue of the imperial past. It also sprang from the feeling that Britain's parliamentary, industrial and cultural achievements embodied a wealth of experience and conferred a moral authority that no other country could match."[75]

In this process, nostalgic-virile leaders (consciously or not) compare their country's actual status, capabilities, and the deference of others to what they were during the "good old days." Reacting with the military to a provocation in the periphery is one way of showing that the key criteria of great power status are being met. Bringing back the "good old days" implies that a country is still the same as it was in the past and can continue to take similar actions. This means reacting to challenges and provocations in the same way as before, and/or using the same national character traits. Showing that the country is as strong as it was previously also means reacting with the same intensity and/or with what can be called the same "tools of power." These are symbols of a country's great power status and how they achieved this status. It can be a capacity, such as diplomacy or a colonial empire. It can be material, such as the armed forces. Or it can even be a type of military equipment, such as aircraft carriers. Challenging a naval power in its periphery, for example, is a symbolic denial of that country's maritime superiority of "ruling the seas." And, of course, a successful reaction would be perfect for the re-establishment of a state to its "true" role (in the eyes of its leaders): that of a great power.

To briefly summarize, my model argues that when faced with a challenge from a less powerful actor in their periphery, leaders of great powers will react by a powerful military response because of their nostalgic-virile understanding of great power status. The aggressor has not only to be punished, but punished severely. This is not necessarily because of the aggression in itself, but mainly because of its symbolic aspects. The consequence of such a challenge is that leaders (as well as their country) then appear weak, uncertain – or even "worse," effeminate, or castrated, and in decline. Through their mental representations of what represents great power status, nostalgic-virile leaders have a tendency to understand any aggression in their periphery by an actor less powerful than them as an international insult (or even as a narcissist insult). Not only is the image that others have of their country at stake if no immediate and strong action is taken in response, but also the image of the leader, whether held by other world leaders or by the domestic opinion of their own citizens. This, in essence, summarizes the logic of nostalgic virility as a decision-making rationality or belief system capable of informing concrete action and practice.

2

Methodology and Research Design

Mixed-Method Approach to Critical Discourse Analysis

My analytical approach to this study is grounded in the widely shared theoretical assumption that it is not only *what* we say that matters, but also *how* we say it. As such, I have employed a mixed-method discourse analysis to systematically review and analyze my datasets. This approach offers a productive way of testing the relevance and applicability of methods and coding questions based on the initial findings, then subsequently revising, fine-tuning, and supplementing those methods in order to improve their accuracy. The work undertaken by Paul Saurette and Kelly Gordon[1] has particularly inspired my approach to this study. This mixed-method approach requires a detailed and often manual qualitative interpretation and coding of every document before it is analyzed quantitatively. Approaching the analyses in this order ensures that the researcher can detect complex, unanticipated, and implicit narratives and rhetorical frameworks within a discursive field and take them into account before "locking" the analysis in with only the predetermined hypotheses and anticipated trends alone. This open-ended process thus ensures that the unforeseen and implicit discursive strategies also appear within and impact the overall findings.

Why is such an approach important? Quantitative methods can answer questions such as "how often" or "how many times" a word, expression, or lexical field is used in any given data. By contrast, with qualitative analysis researchers can unpack motives, feelings, and perceptions – those revealed in discourse through, for example, the choice of words and emotive expressions. Scholars have recognized that choices of language and vocabulary play an important role when it comes to both expressing and constructing perceptions, emotions, and identities, as well as identifying them in discourses.[2] Qualitative analysis allows us to uncover these emotions, perceptions, identities, and a common frame of reference for them.[3] A close qualitative reading of the data thus helps uncover the persuasive mechanisms and analogies at work in discourses that quantitative methods alone are unable to detect. It enables the researcher to identify complex rhetorical formations and verify if the language used is coherent with the type of document under analysis (or across different types), and – even if it is not – what analytical insights this reveals.

However, qualitative analysis alone does not give a sense of how broadly the discursive field is impacted or characterized by any given trend, or how frequently specific mechanisms are invoked therein. Accordingly, I also use a quantitative approach to measure the frequency of idioms, narratives, analogies, and words. The basic assumption is that the more a word or rhetorical strategy is used, the more important its role within the broader discursive field. This enables us to clearly identify the rate of both their occurrences and what Max Reinert calls co-occurrences – that is to say, the words/ideas used together in the same semantic unit.[4] However, quantitative analysis provides only a quantification of empirical research materials; it is then the researcher's task to interpret the numerical results in a cultural and historical context through qualitative analysis.[5] In this sense, the mixed-method approach comes full circle to mutually strengthen each element of the analysis.

Before presenting the details of my dataset construction and composition for the three case studies presented in this book, I will first discuss in more detail the empirical research design, choices, and processes involved in my mixed-method discourse analysis.

Analytical Process and Interpretative Method: Coding Protocol and Coding Dictionary

Defining a Coding Protocol and Constructing a Coding Dictionary

It must be stressed that my systematic mixed-method discourse analysis is *not* an "automated" content analysis (e.g., using computers to generate word counts), but quite the opposite. The difference between my mixed-method approach and a traditional qualitative approach is that I also systematically direct and track the qualitative interpretations and readings with the help of a coding dictionary and with the statistical functionalities of the QDA Miner software. This sort of tracking could be done simply with coloured pencils and a basic calculator, but using QDA Miner allows for a much more user-friendly, efficient, and accurate tracking and quantitative perspective on the close qualitative readings.

To ensure that the most reliable, valid results are provided, and that all documents (for all case studies) are coded in a consistent manner, I followed a three-stage coding protocol inspired by Halperin & Heath[6]:

Stage 1 – Design themes for a coding dictionary to be used as a framework for qualitative coding of all datasets. The codes were created based on the literature and each aspect of the theoretical frameworks I wanted to test. Building a coding dictionary is not just about creating codes; it is also, and essentially, about identifying indicators that have to be found in the data in order for the code to be applied.

Stage 2 – Test the coding dictionary with a representative sample of the dataset. The objectives of doing so are (a) to verify whether there was a need for other codes or if the codes are operationally efficient/relevant, and (b) to generate inductive hypotheses that were then translated into coding questions before being transformed into hypotheses and added to the theoretical frameworks; this was to ensure that no relevant variants of the theoretical frameworks were omitted.

Stage 3 – Code the documents. Coding a text means applying a tag for each variable's category and for each subcategory. When a passage is read, it is compared to the coding dictionary and – depending on how close it is to the dictionary and to previous passages already coded, it is then coded using the same or a different code, as appropriate. The use of qualitative data analysis software such as QDA Miner is invaluable because multiple codes can be used for the same words/sentences/paragraphs, and data associated to each code can be quickly retrieved. Coding was checked several times to ensure consistency throughout.

Clear, robust conceptualizations and coding definitions, along with the ability provided by QDA Miner to systematically keep track of every element identified during the coding, help address an important risk that researchers face: namely, seeing only what the theories being tested assume.[7] Use of QDA Miner and a strict adherence to the coding protocol help ensure that each theory is treated equally, that unforeseen codes and trends can also be included in the analysis, and that the results are not biased in favour of any given theoretical framework.

Dataset

Data Collection Principles

To conduct this study the choice of data was key, especially as there are significant risks to any study of the perceptions and emotions of individuals.[8] These risks can include: (i) uncertainty over whether what is said or written by leaders is "the truth and nothing but the truth"; and (ii) the possibility of "hidden interests," wherein the ideas, emotions, identities, and perceptions expressed might merely be a useful mask to hide other material interests.[9] One way of overcoming this is through the use of declassified documents, which, theoretically at least, limit the presence of these risks within a dataset by gaining access to otherwise private or publicly withheld discourses.

Using declassified documents does not automatically guarantee absolute truth-value or sincerity of the discourses therein. Nonetheless, building my claims from the analysis of declassified documents helps me to avoid the bulk of the two risks mentioned above. These documents are, by definition, designed to remain confidential, meaning their content was not written to please domestic opinion and that there is no need for decision-makers to "sell their arguments" within them, as is the case with public speeches.[10] The datasets for each case study consist of documents that were highly confidential, even "secret," at the time of their respective events. They consist mostly of private letters between decision-makers, cabinet minutes, war committee reports, notes from intelligence agencies, reports from embassies, and transcripts of meetings between decision-makers. Some documents have an even greater significance because they have been annotated by leaders themselves, thus giving us very precise insights into what they were thinking about the various options they had at the time.

Although my datasets consist of primary sources, to support and help interpret them I also draw on secondary sources and existing literature, such as historical studies. These are often useful in reconstructing a relatively concise overview of the most important events, positions, trends, and discursive tendencies prior to the case being analyzed. I also use these secondary sources to narrow the time frame of the analysis of each case study as well as to identify key actors and the most commonly recognized arguments or positions. In respect of the significance of certain words: depending on the context and date, the meaning of words can change in important ways. These changing definitions or connotations – which can have important implications for interpreting historical discourses – will not be found in declassified documents but rather in secondary sources.

In terms of data collection sources, my datasets were constructed using primary sources from various archives in France, the UK, and the US (more specific details on the composition of each are included below). Once in the archives, I followed the collection method suggested by Una Bergmane.[11] The decision on whether to include documents in the dataset was based first on their direct or indirect relevance to the case studies. After this first selection, a second selection process was carried out based on a second criteria: originality. This criterion was used to eliminate unnecessary documents which could have potentially skewed the results of the statistical analyses by giving more importance to some themes or trends than they deserved (given their relative lack of relevancy or importance to the discursive fields in question). Second copies of documents were not selected unless there were changes from the original, or if there were varying copies sent to different audiences (e.g., the original document to embassies and an amended copy for MPs).

Dataset Management

For better management of the data and to offer some more nuanced analytical comparisons, the three datasets were each categorized and studied based on several "filters" (or "variables" in the QDA Miner vocabulary). In total, I used three types of filters:

(1) The *Phase* variable. Each conflict was divided into three phases. The exact chronological distinctions and timeframes are outlined below. Generally speaking, however, *Phase 1* is used as a neutral comparison point as it goes from the first entry in the archives to one day before the provocation. Doing so allowed me to interpret how the issue was seen by decision-makers before the conflict, helping me to better measure the impact of the provocation on the decision-making process. *Phase 2* corresponds to the day of the

provocation until the day the order of intervention was given to the military/departure of the troops – a critical moment when the decision to go to war is made. Narratives and arguments found here can, therefore, be considered as qualitatively more important than those found at the two other phases (although not unaffected by them). *Phase 3* goes from the first day after the order of intervention is given/departure of the troops until the end of the conflict. This phase helps to identify which narratives and arguments were given to allies and domestic opinion to justify the decision.

(2) The *Actor* variable. By this, I refer to the fact that the collected documents have been distributed based on the different actors involved: heads of state/government; ministers, advisors, and MPs; military; and domestic opinion. This variable offers valuable comparisons based on what each group of actors argued.

(3) The *Type of Document* variable. For practical reasons, I made only three divisions: (i) *internal (government) documents*, comprising all documents that were meant to remain within the government and never be released (e.g., cabinet minutes, memos from advisors to heads of state/government, proposal drafts, transcripts of cabinet discussions, correspondence between various government units, military reports, etc.); (ii) *external (government) documents*, comprising any documents that were to be sent to other governments, such as letters from one head of state to another, or transcripts of meetings or phone calls between decision-makers from two countries; and (iii) *public documents*, comprising those documents that aimed from the start to be released to the public, such as interviews, public speeches, and letters in response to MP's questions. Based on this division, it is assumed that the narratives and arguments found in *internal documents* will be franker and more credible than those found in *public documents*, because of their private or classified nature.

Dataset Composition

FRANCE

Data was collected from the National Archives (Archives Nationales), Diplomatic Archives (Archives Diplomatiques), Ministry of Defense Archives (Service Historique de la Défense), and the National Archives of

Table 2.1 Distribution of French documents based on actors

Actor	*Number of documents*
Head of government	9
Military	77
Government	54
High commissioner in Indochina	52
MP	105
Press	30
Total	*327*

Overseas Territories (Archives Nationales d'Outre-Mer), as well as from the French National Assembly and Senate websites (for parliamentary debates). The French dataset is composed of 327 documents. Unfortunately, the dataset is not complete due to the partial composition of some archive files.[12] They nonetheless offer ample materials for a robust and systematic analysis.

For better management of the data, and to strengthen the analysis and comparisons, I categorized the dataset based on the three chronological phases:

Phase 1: 24 March 1945 to 18 December 1946. The first date corresponds to de Gaulle's speech in which he highlights French objectives and strategy regarding Indochina. The second corresponds to the day before the Viet Minh attack in Hanoi, considered by leading historians as the real start of the First Indochina War.

Phase 2: 19 December 1946 to 22 March 1947. The first date is the day of the Viet Minh general attack against the French in Hanoi. The second corresponds to the vote of war credits for Indochina.

Phase 3: 23 March 1947 to 23 December 1947. With the Indochina War being relatively long, it was difficult to decide which date should be used to mark the end of the data collection period. This phase represents the time during which the government "sold" the war to the French domestic audience and chose whether or not to reveal the narratives that had made them decide to go to war. I chose to stop

Table 2.2 Distribution of French documents based on the phase of the conflict

Phase	*Number of documents*
1	147
2	77
3	103

data collection at 23 December 1947 because it was the date that any negotiations with the Viet Minh have been *de facto* officially ruled out.

BRITAIN

The British government began declassifying documents related to the Falklands crisis in 2010. These documents relate to the negotiations with Argentina and cover the crucial period from 1979 to 1982, when the situation spiralled out of control and ended up in a military conflict. They are largely comprised of reports from the cabinet's internal discussions, communications from the Falklands governor, and private letters from Thatcher and her ministers to other heads of states, such as Ronald Reagan.

All the data comes from the Thatcher Foundation website (it can also be found on the British National Archives website). In total, 1,840 documents related to the Falklands War have been declassified and published by the British government. I have limited my dataset to 361 of these documents, dating from 13 May 1979 (Thatcher became prime minister on 4 May) to 14 June 1982 (the day Argentina surrendered). I selected only those documents that have a clear link to the conflict. Neither copies of documents nor documents reporting military operations have been selected, as the objective is to explain the *reasons* that triggered the British reaction rather than the logistical details of the way in which Britain conducted the conflict.

However, the tables below show slightly more than 361 documents. The reason is that some documents are cabinet minutes and recordings of conversations that contain content spoken by ministers as well as Thatcher. In these instances, I broke these documents down into sub-documents for the purposes of categorization (e.g., creating one for ministers, one for Thatcher, and one with content that could not

Table 2.3 Distribution of British documents based on actors

Actor	*Number of documents*
Thatcher	120
Cabinet	27
Ministers	138
UK Embassies	26
Military	36
Falklands Government and officials	19
MPS	28
Other (mass media, opinion polls, unidentified)	19
Total	*413*

Table 2.4 Distribution of British documents based on the phase of the conflict

Phase	*Number of documents*
1	145
2	91
3	177

be attributed to any one individual and has been classified under "cabinet"). In other instances, I had to create documents to represent, for example, those where Thatcher wrote comments on the margins, in order to distinguish between the official transcriptions or letter and the extra commentary.

As with the French dataset, for better management and analysis of the data the conflict has been divided into three chronological phases:

Phase 1: 13 May 1979 (Thatcher's ascent to power) to 18 March 1982 (the day before the Davidoff incident). *Phase 1* will be used as a parameter to measure frequencies before the conflict.

Phase 2: 19 March 1982 (the day of the Davidoff incident) to 5 April 1982 (the departure of the British task force). *Phase 2* corresponds to the moment the decision to go to war was taken, and where the real motives are expected to be most present.

Table 2.5 Distribution of US documents based on actors

Actor	*Number of documents*
Reagan	42
Advisors	47
Government	110
US Embassies	69
Military, CIA, National Security Council	52
Press	37
MPs and Congress	9
Public	24
Total	*390*

Phase 3: 6 April 1982 (one day after the departure of the British task force) to 14 June 1982 (Argentine's capitulation). *Phase 3* corresponds to the moment the decision was taken to send the fleet when the conflict started and when it came to an end, as well as the period of "selling the war" to the UK domestic audience.

UNITED STATES

In total, 390 documents were collected from CIA archives and from the Ronald Reagan Library. Data collection for this case study was rather frustrating for several reasons. In some cases, documents had no date or author, adding significantly to the amount of work required to acquire such information. When this proved impossible, or when identification was judged unlikely by the foundation archivist, documents were not included in the dataset. Tables 2.5 and 2.6 show the distribution and organization of the coded dataset.

To clarify, the "public" category was composed of letters sent to Reagan and his administration by ordinary US citizens. In the case studies, the documents are cited based on the files in which they appear in the archives. In other words, it is the name of the person to whom the files belong that is given in the archives, and in some cases, this differs from the name of the author of the actual document. When the author was indicated on the document, it will be specified in the reference.

Table 2.6 Distribution of US documents based on the phase of the conflict

Phase	*Number of documents*
1	159
2	149
3	82

One important note to discuss here is the diversity and difference between the types and quantities of documents available for each case study. While it might appear to some as a limitation in terms of comparability, I argue that the diversity of documents and contexts captured by these datasets serves to reinforce our findings, given that the insight into different contexts and sources of discourse would demonstrate, where positive results are found, the empirical reliability of each tested framework or prediction across diverse contexts and situations.

As with the two previous case studies, the dataset has been organized into three phases. This proved difficult in terms of data categorization given that, in some cases, there was no chronological indication or dates on the documents, meaning that I had to look for indications within the text to be able to organize the dataset chronologically. When no indication could be found, the document was not included. The three phases are as follows:

Phase 1: 1 January 1980 to 18 October 1983. This stretched from the start of Reagan's presidency to the day before the coup in Grenada that overthrew Bishop. The length of this phase is for a simple reason: the literature claims that Reagan wanted to get rid of the communist regime in Grenada from the outset of his presidency. To investigate such a claim, it was necessary to track Reagan's thinking about Grenada from the start.

Phase 2: 19 October to 24 October 1983. This phase corresponds to the day of the coup in Grenada to the day before US troops landed in Grenada. During this phase, the decision to invade Grenada was taken.

Phase 3: 25 October to 31 December 1983. This phase covers the fighting and the immediate aftermath of the war.

Supplementary Cases[13]

While they are not presented in great detail, chapter 9 concludes by summarizing the findings of two supplementary studies of more contemporary conflicts using the same theoretical framework and coding dictionary: namely, the Russian conflict in Ukraine and the South China Sea dispute. These studies have been published and presented elsewhere, but for the sake of clarity I will briefly outline the dataset composition for each study here.

CHINA

In total, 365 official public documents have been collected, not without difficulty (due to access restrictions). These documents consist mostly of transcripts from press conferences held by the Chinese Ministry of Foreign Affairs and the Ministry of Defence, as well as speeches made by either ambassadors, ministers, or President Xi Jinping. It also comprises Chinese legal documents produced at the occasion and in response to the development of the Permanent Court of Arbitration sentence on 16 July 2016. To ensure the possibility of comparing these diverse discourses (and the specific words and rhetoric used therein), they have all been collected in their official English translations.

For better management of the data and to offer more nuanced analytical comparisons, the dataset was categorized and studied based on three key phases of the issue:

Phase 1: stretches from 4 February 2014 to 28 June 2016, excluding the first question asked to Chinese officials about the sentence. Phase 1 is used as a neutral comparison point. We will thus be able to see how the issue was viewed by Chinese officials, which will help us better measure the impact of the sentence on Chinese perceptions and narratives.

Phase 2: stretches from 28 June 2016 (inclusive) until 8 September 2016 (the end of the twenty-eighth ASEAN summit in Laos, during which there is an agreement not to mention the ruling). This phase constitutes a time of important tensions and rigorous hostile Chinese discourses opposing the procedure that led to the ruling and the content thereof.

Phase 3: stretches from 9 September 2016 to 30 August 2020 (the last entry in the dataset). This phase will help us identify which

Table 2.7 Distribution of Chinese documents based on the phase of the issue

Phase	*Number of documents*
1	182
2	34
3	149

Table 2.8 Distribution of Chinese documents based on actors

Actor	*Number of documents*
Ambassadors	25
Foreign ministry spokesperson	296
Ministers	60
Xi Jinping	5
Total	*386*

narratives are used by Chinese officials to justify their refusal to enforce the sentence.

RUSSIA

In total, seventy-nine official public documents have been collected from 2012 to 2023, with great difficulties to access Russian official websites because of the war in Ukraine. These documents consist mostly of transcripts from press conferences, interviews, and public speeches held by Dmitri Medvedev, Vladimir Putin, and Sergei Lavrov. To ensure the possibility of comparing these diverse discourses (and the specific words and rhetoric used therein), they have all been collected in their official English translations.

For better management of the data and to offer more nuanced analytical comparisons, the dataset was categorized and studied based on three key phases of the issue:

Phase 1: stretches from 1 October 2012, first entry available, to 9 April 2021, the day prior the beginning of the Russian-Ukrainian

Table 2.9 Distribution of Russian documents based on the phase of the conflict

Phase	*Number of documents*
1	52
2	15
3	12

Table 2.10 Distribution of Russian documents based on actors

Actor	*Number of documents*
Foreign ministry spokesperson	11
Dmitri Medvedev	29
Sergei Lavrov	19
Vladimir Putin	19
Total	*78*

diplomatic crisis of 2021–22. Phase 1 is used as a neutral comparison point.

Phase 2: stretches from 10 April 2021 (the beginning of the diplomatic crisis between Russia and Ukraine) until its culmination in the invasion of Ukraine on 24 February 2022. This phase constitutes a time of important tensions and more hostile Russian discourses.

Phase 3: stretches from 25 February 2022 to 21 February 2023 (the latest public speech of Vladimir Putin at the time this book was written). This phase will help us identify which narratives are used by Russian officials to justify the war and cope with international pressure and casualties.

Limitations

A few potential limitations of this study and methodology should be briefly mentioned. For instance, at the level of methodology and research design, there will be results that critical discourse analysis alone cannot obtain. Namely, it cannot always explain why one narrative was used and not another, why a specific idea was present or absent within a narrative, etc. There is no absolute guarantee of the "real" motivations of the actors, what they actually truly believed – although no approach can dispel this limitation. As well, the results obtained in my case studies cannot be completely reliable due simply to the nature of the datasets used. Indeed, as this study used almost exclusively declassified documents (with some public documents added afterward for capturing different perspectives), the results could change if other documents were to be declassified in the future (e.g., in the US case study, the results can only be viewed as temporary as many key documents are still classified). Alternatively, the results could be skewed in certain ways simply because of the nature of the archives (e.g., in the French case study, the archives are clearly incomplete).

Finally, at the level of theorization and generalizability of the model forwarded here, this study offers a novel theoretical framework for an ideal type of status-seeking and nostalgic virility among leaders, the combination of which should be understood as probabilistic rather than deterministic. What I hypothesize is that, all things considered, it is likely that great powers go to war in their periphery against less powerful actors because of a combination of these two factors. Although it is probable that most individuals will not fall neatly into these categories and will not match all the criteria identified, those who do will display signs and tendencies accordingly that push them towards the decision or rationalization for going to war. Further research should be done to see if nostalgic virility plays a role in other status and power distribution contexts. In other words, the status-seeking and nostalgic virility framework should be viewed as a theoretical and explanatory framework in its infancy; it will require further investigation and nuance in order to fully understand and contextualize its role within the IR landscape.

Having laid out the theoretical framework and research design and methodology in detail, let's now turn to an examination of the findings for each case study under analysis.

3

France in Pursuit of Past Glory in Indochina, 1945–47 (I)

Testing Mainstream Theories

The First Indochina War cannot be understood without reference to the consequences of the Second World War. France was originally defeated by Nazi Germany early in the conflict, yet managed to appear as one of the "victors" through the illusion presented by Charles de Gaulle's *France Libre* movement. On the ground, France was materially closer to ruin than to victory. The country was in need of reconstruction not only from a material point of view, but from a political and moral one. After the Vichy collaboration with the Nazis, a new political system was required. The question of re-establishing French authority in the colonies was also on the agenda, especially in Asia where Japan had defeated France and removed the French colonial system in 1945, just before capitulating to the US.

The French attempt to re-establish authority in Indochina proved to be rather difficult and ultimately led to a decolonization war after 1947. As various military and historical studies have shown, the processes and events that pre-empted and led to the war were complex: interaction with post-Second World War diplomacy and the division of Indochina into various occupation zones between the UK and China; the will of the former colonial power, France, to regain control over its colony despite US hostility; the influence of the early phases of the Cold War; and the beginning of the organization of independence movements in European colonies. It is relatively surprising that France decided to try and re-establish control of Indochina immediately after the Second World War, at a time when both its own economy and political system had to be reconstructed and when France clearly lacked the financial means to do so. Historians have suggested various dates for the precise start of the Indochina Wars, but this study concentrates on the early phases: from the French landing in Indochina (1945) until the end of 1947, which marked the point of no return.[1]

A Panorama of Existing Literature

Plenty of literature has been written about the First Indochina War, especially on military aspects such as the operations themselves and the use of specific tactics.[2] The conflict has also been examined by political analysts and constitutional law scholars looking at the effects of the war on the institutions of the French Fourth Republic, on relations between different French political parties,[3] and on the collapse of the French colonial empire.[4]

However, most of the literature consists of detailed studies about the origins and development of the conflict.[5] Historians have presented several arguments for why France chose to go to war, some of the most prominent being: (i) France was pursuing the liberation of French territory, from the metropolis to the colonies; (ii) they sought to recoup or revive French status; and (iii) they sought to re-enforce colonial rule and thus protect the population from attacks by the Viet Minh. Historians have also shown how the personal agendas of French local authorities may have impacted events – for example that individuals such as General Jean Valluy and High Commissioner Thierry d'Argenlieu, through their inflexibility, directly impacted the course of events and probably sped up the march to war.

A Priori Evidence to Support the Three Tested Theoretical Frameworks

In this section, I will present a priori evidence to show that the three theoretical frameworks being tested in this study are relevant to any explanation of France's decision to go to war. This evidence is taken from the current literature and constitutes a starting point for analysis.

Status and Nostalgic Virility: A Priori Evidence

On 10 May 1940, Germany launched an offensive against France; the war lasted only six weeks.[6] On 16 June, an armistice was negotiated, and France was occupied. This defeat had serious consequences for France's position on the world stage.[7]

To make the situation even worse, during the Second World War France had been forced by Japan to make ever more concessions, and three provinces belonging to Cambodia had to be transferred to the Kingdom of Siam (Thailand). Things worsened even further on 9 March 1945 when Japan organized a coup to get rid of the French military and administration. Soldiers and officials were imprisoned or massacred. Japan then proclaimed the independence of the countries composing Indochina.[8] These events proved very damaging to French status, especially as the underlying structure of the colonial system rested on the racial (and virile) superiority of the "white man."[9]

Not only were the French defeated, they were also publicly humiliated by the Japanese who paraded them on the streets before the Indigenous people who they were supposed to have been able to protect due to their "racial superiority."[10] As a result, the inferiority complex on which the colonial system was based suddenly disappeared. In the face of these losses and failures, and with colonial rule being challenged elsewhere (e.g., Tunisia, Syria, Madagascar), giving up on Indochina could have risked a complete collapse of the colonial empire.

Initially after the Second World War, France did not even have an occupation zone in Germany or a seat on the United Nations Security Council, and was granted both only because of British pressure on the US. In this context, the idea of giving up their colonies was perceived as a show of weakness. An IFOP (Institut Français d'Opinion Publique) survey dated 1 July 1945 showed that 80 per cent of French people were convinced that their country was still a great power.[11]

Nostalgic virility assumes that leaders compare their country's current world status with that of the past, particularly in times of glory. Pre-1940 France and post-1945 France were worlds apart, and many historians agree that nostalgia – especially that of French citizens living in Indochina who saw their status and dominant position being challenged on an everyday basis – played a role in the interaction between actors in Indochina.[12] Therefore, any challenge posed to French possession rights over Indochina may well have been used as an opportunity for France to ensure that its 1940 military defeat (and subsequent challenges) would be forgotten through the display of effective military force.

Realist-Strategic Frameworks: Securing the Colonies to Reconstruct France?

A priori evidence supporting the framework rests on the situation in which France found itself after the Second World War. Jane Burbank and Frederic Cooper argue that both France and the UK needed their colonial empires after the war because they had seen their industries destroyed and had had to borrow a huge amount of money from the US.[13]

Sales of exotic colonial products were seen as the only means for these countries to earn the foreign currencies (especially US dollars) required for industrial reconstruction and to regain their lost global influence.[14] A colonial empire was also required from a strategic and military point of view. It is important to remember that the colonies had played a key military role by providing troops during the Second World War. In fact, most of the *France Libre* military forces came from colonies that refused to follow the Vichy regime. Losing colonies meant losing military power, not to mention potential key strategic locations.[15] Accordingly, there were clear potential motivators for a material or realist-strategic interpretation of France's decision to go to war.

DTW Frameworks: Stabilizing Governments?

Finding a priori evidence in support of the DTW framework in the context of French politics between 1945 and 1947 is rather difficult, as there was no firmly established political system. French politics at the time were quite unconventional: it was a period of transition requiring new institutions after the period of Vichy rule.

The Constitution of the Fourth Republic, although approved by referendum on 27 October 1946, was not applied until the election of President Vincent Auriol on 16 January 1947. Until that date, France was under a provisional government in opposition to many projects for a new constitution. How might the DTW framework be expressed in this context?

One possibility is that there was a desire to gain public support in favour of one of the constitutional projects, or for upcoming parliamentary elections once the new constitution was finally approved (possibly by looking at the success rate of governments). There had been four heads of government from different political parties and movements between 1944 and 1947: Charles de Gaulle (June 1944–January 1946), Félix Gouin (January 1946–June 1946), Georges Bidault (June 1946–November 1946), and Léon Blum (December 1946–January 1947). So, one possible way to see the DTW framework in action would have been through the war in Indochina: it may have been seen as a way to either gain or consolidate a majority in parliament.

As this brief review of a priori evidence for the three theoretical frameworks under analysis demonstrates, all three options pose possibilities for explaining the French decision to go to war in Indochina. In the remainder of this chapter, I will test these assumptions against the empirical data to determine the degree to which decision-makers' motives and perspectives embody these theoretical and a priori explanations. Evidence for both the realist-strategic and the DTW frameworks is presented in this chapter, followed by a presentation of the evidence for the status and nostalgic virility frameworks in the next chapter. Throughout each section, efforts are made to put these theoretical frameworks into conversation with one another based on concrete findings from the mixed methods analysis of declassified French documents.

Testing the Realist-Strategic Framework: Very Little Evidence

The Surprising Absence of Strategic-Material Arguments in Discussions of Whether to Go to War

Overall, mentions of strategic-material ideas are found in some 52.6 per cent of documents. This high score might lead to the initial conclusion that strategic-material reasoning was key in the decision to go to war. However, this would be incorrect. On digging deeper, we find that while strategic-material ideas were found in many documents, the arguments were primarily incidental and logistical. Indeed, explicit mentions of going to war *because of and for strategic-material reasons* were found in only 2.8 per cent of documents. As I explore in further detail below, the mere use of such strategic-material terminology in 52.6 per cent of the cases cannot be translated into the expression of a strategic-material *justification* for going to war.

Using Indochina to Reconstruct France? Economic Interests Marginalized to Specific Situations and Contexts

Given the situation in which France found itself at the end of the Second World War, I expected to find a strong argument based on the idea that re-establishing French authority and rule over Indochina would provide leverage for reconstruction. In total, economic arguments were found in 2.1 per cent of cases overall. By retrieving all instances of the codes relating to economic interests, it appears they were concentrated in *Phase 1* (2.1 per cent, or four instances) and *Phase 2* (3.9 per cent, or three instances), and nearly disappeared in *Phase 3* (1 per cent, or two instances).

The arguments identified centred around two codes: *potential trade gain/loss* and *potential gain/loss of natural resources.* From these, two distinct narratives were identified. The first was found in *internal documents* (those issued from within the government apparatus and aimed to remain within it, rather than rendered public) at *Phase 1* – the moment when France sent an expedition nary force under the command of General Leclerc to re-establish French control over Indochina. It was constructed around the idea that in order to increase economic gains, it was in France's interest to grant Ho Chi Minh the independence he requested for Vietnam. As Jean Cédile[16] reported to Leclerc: "They [the Annamite government] have largely insisted on the fact that a simple declaration of independence would bring the Indochinese people closer to France … Our country would keep its privileged and largely dominant positions. Indeed, the Annamites envisage the signature of cultural, economic, diplomatic and military agreements, which would leave us control over the future of Indochina."[17]

The second narrative had the opposite goal in mind: staying in Indochina. This was constructed from two ideas that appeared in 1947 (at the end of *Phase 2*). The first idea focused on the benefits for France of remaining in Indochina, especially to help in the reconstruction of the country. It was essentially found in parliamentary debates, such as those during the 18 March session.[18] Using Indochina's resources to reconstruct France seemed to be expected from an imperial point of view: the colonies had to support the metropolis, and France was in dire straits.[19]

The second idea used in the economic narrative supporting French involvement links the economy to symbolism and emotions: Indochina could develop itself only because of French involvement, and now that it was a colony (finally) providing gains, it would be a bad mistake to abandon it. This narrative was also found in parliamentary debates: "[The French] would be even more surprised if they were told that in a few decades, the French transformed Indochina into the biggest rice exporter in the world after Burma!"[20]

On the whole, the analysis of French documents shows that economic interests are very unlikely to count as strong motives for the French involvement in Indochina. Throughout the conflict, no strong measure of evidence of economic concerns was identified from key decision-makers, in either quantitative or qualitative terms. Accordingly, we will now consider whether any strategic or geopolitical concerns proved to be motivating factors in the conflict.

From Avoiding the Collapse of the Empire to an Initiated Cold War Conflict: Various Strategic Frameworks Tested Lacked the Support of Empirical Evidence

Among the many codes specifically designed to measure direct strategic and geopolitical concerns, only three were found in the dataset: *potential gain/loss of a strategic location*, *fear of a domino effect*, and *potential gain/loss of a bridgehead*. Overall, these codes were found very infrequently in documents (1.5 per cent, 0.6 per cent, and 0.3 per cent respectively). Their frequency was very low in *Phase 1* (below 2 per cent) and completely absent in *Phase 3*. Due to their increased frequency in *Phase 2*, however, I assumed that there may be strong qualitative evidence to compensate the relatively low frequencies. The results obtained were mixed at best.

The French military had assessed the strategic importance of Indochina in rather ambiguous terms. As early as May 1945, General Sabattier, de Gaulle's personal representative in Tonkin, wrote in a note to him that "France has very important interests in Indochina." Yet, he does not mention what the interests were, only that it was because of them that China was not very keen on seeing France re-establish its position there.[21]

Explicit mention of strategic interests associated with Indochina were found, but much later in the dataset, in documents dating from late 1946 and into 1947. It appears, rather, that strategic interests were strongly associated with status considerations, and focused notably on the definitions of great power. This will be further detailed in chapter 4.

For now, I will concentrate on investigating the precise strategic interests that *were* mentioned – namely, safeguarding the integrity of the French colonial empire. It was felt that failing to re-establish French rule in Indochina would result in the collapse of French possessions elsewhere. As the newspaper *Résistance Paris-matin* noted: "The failure of France in the Far East would be followed by the same failure in all our overseas possessions."[22] This type of argument was also made by some MPs.[23]

The President of the Provisionary Government, Bidault, clearly identified the possessions that were seen as threatened by the situation in Indochina when noting in a November 1946 session of the Indochina Committee (specifically tasked with managing the situation in Indochina, and abolished on 8 January 1947) that "nothing should be done in Indochina that could be used as a precedent which could later be used in respect of the situation in Morocco and Tunisia, nor in terms of concessions, nor on the aspect of invitations."[24] North Africa was a key part of the second French colonial empire built in the nineteenth century and, at this time, the situation there was rather critical: the navy and air force had both been used to support ground forces in order to regain control of the situation.[25]

Given the scant evidence for strategic and geopolitical interests, another avenue worth examining is the potential influence of the Cold War. Were there any connections between this geopolitical context and France's strategic-material interests? Was there an East-West divide obvious in French discourses related to Indochina? After all, the conflict clearly took a Cold War-related turn after 1950: was this already apparent between 1945 and 1947?

From a quantitative point of view, the results are indecisive because the frequencies are too low. In the entire dataset, the code *fight against communism* was found in only 2.1 per cent of cases in *Phase 1*, growing to 2.6 per cent in *Phase 2*, and decreasing to 2 per cent in *Phase 3*. Nor, for various reasons, was the evidence decisive from a qualitative point of view. At the time, the French Communist Party was still part of the governmental coalition. As such, evidence retrieved from the dataset is mostly centred on the description of some individuals as being communist.

The first mention, in 1945, was in a military report explaining that China's presence in Tonkin was due to its willingness to crush the Viet Minh because of their communist affiliations.[26] Later French documents reveal that it was a question of whether to grant authorization for a communist political group in Indochina to publish a newspaper,[27] and also that in Indochina nationalism was synonymous with communism.[28] Mentions of communism were essentially found in records of parliamentary debates (3 per cent), while available government documents prove very silent about the fact that the Viet Minh were influenced by communism. Other than General Juin's previously mentioned report, the only reference is in a document from General Valluy (supreme commander of French forces in Indochina), who simply noted that "Ho Chi Minh has fully embraced radical communism."[29]

At the outset, these two strategic narratives appeared to be good explanations for the French involvement. However, analyses of the dataset proved that this was not the case, at least for the period 1945–47. Indeed, this time period was probably too early to see any Cold War narratives emerge: French communists were still participating in the government and the East-West divide, formulated in Churchill's 1946 "Sinews of Peace" speech, was yet to be integrated into the public mind. It was only at the occasion of the Korean conflict that Cold War narratives were added to French official discourses.[30]

Interpreting the Quantitative Results: Strategic Concerns Linked to the Conduct of Military Operations and Not the Decision to go to War

In this section I will concentrate on unpacking the strategic-material discourses present in French documents yet seemingly unrelated to the decision to go to war. While there is evidence of strategic-material terminology, ideas, or reasoning in 52.6 per cent of documents, only 2.8 per cent translate into discussions of military intervention in Indochina, by and large found within debates on the cost of military operations in Indochina; requests made by the military for more and new equipment, budget, and troops; strategies for adopting a more efficient form of fighting against the Viet Minh; and within reports of combat (which included casualties).

The cost of the war in Indochina was debated in both Houses of Parliament (20.8 per cent of parliamentary documents). Criticism came mainly from MPs leaning towards the left/extreme left who argued that the war in Indochina diverted money that could be better spent on financing social programs.[31] On the opposite side was the military, which claimed they lacked the appropriate budget to fight in Indochina. They requested more men, additional equipment, and a renewal of what they already had. For example, in January 1947, Leclerc wrote in the report following his inspection mission in Indochina: "The worst would of course be to let the expeditionary corps weaken, which would be equal to progressively giving up Indochina. The best would be to increase it by putting its number to 115,000 men."[32]

With regard to the issue of equipment, Valluy, in charge of operations, asked d'Argenlieu to get boats.[33] Renovating existing equipment that had been provided by the US in the Second World War was vital but problematic – as acknowledged during the Defence Committee on 3 May 1947: "Most of the American equipment is worn by three years of war and is working only because of an experienced staff who will soon be recalled and replaced by a staff essentially composed of North-Africans. We fear that the equipment will soon be out of use."[34] These concerns were found countless times but never linked to the realist-strategic framework as an explanatory factor for going to war.

Conclusions on the Realist-Strategic Framework

What conclusions can be made about the relative importance and impact of the material and strategic frameworks on the French decision to go to war in Indochina? Although codes capturing strategic and material issues were found in many cases, they in fact captured concerns linked more to the logistical conduct of hostilities (costs, need for more supplies, men, and equipment, what tactics to use, etc.) and not to the decision of whether or not to go to war.

In regard to the narratives linked to the realist-strategic framework, we find that the importance of natural resources was mentioned and taken into account, but both quantitative and qualitative analysis shows that it was very unlikely they provided a motive to go to war. Rather, codes capturing the expression or relevance of the realist-strategic theoretical framework were mainly centred around the strategic location of Indochina. This meant that events in Indochina were perceived as having consequences for other parts of the French empire, and that France did not want to risk what was internally seen as a possible domino effect for their other colonies. Nonetheless, the minor role that this narrative played in the overall dataset displays that this too was unlikely to be a primary motivator in the decision to go to war.

Finally, codes measuring the impact of the Cold War on French decision-making were also infrequent and qualitatively weak. Where communism was mentioned, it was mostly to describe the political views of the Viet Minh and not in the context of geopolitical East-West competition. Most probably, it was simply too soon to find such concerns, which began to develop in earnest after 1950.

Overall, then, we find that realist-strategic concerns were minimal in French decision-making and discourses on the situation in Indochina. Based on the mixed-method analysis of the available data, we may conclude that the French decision to go to war in Indochina was not in any strong way related to the realist-strategic framework of assumptions. So, what may have motivated French military action? The remainder of this chapter will outline and evaluate the evidence for the diversionary theory of war to determine whether or not domestic political factors played a decisive role.

Evaluating the Diversionary Theory of War in Indochina: An Exotic War to Gain Political Support? No Empirical Support at All

Quantitative Results Backed by Polls: The French Were Not Interested in Indochina

To measure and analyze evidence for the DTW framework, I have included a diverse set of codes to capture as many expressions of the framework as possible. The coding dictionary was thus organized around four main categories: (1) Measuring any *fear of losing power*, which also encompassed the ideas of losing party leadership or a government/ parliamentary majority; (2) Measuring any *fear of losing upcoming elections*, which also encompassed codes targeting explicit or implicit mention of any type of election to be won or feared to be lost (e.g., general, parliamentary, party, local); (3) Measuring any *fear of losing the support of domestic opinion* (e.g., concerns about polls, influence of lobbies, potential strikes, party opinion, parliamentary hostility, etc.); and (4) Measuring the DTW framework in relation to any explicit or implicit reference to the *need to divert domestic opinion*.

As mentioned, France's domestic situation after 1945 was critical. Historians report French frustration, especially over the fact that they still had to use ration tickets to manage low supplies.[35] Because of this, I was expecting to see an association between the domestic situation and the idea of intervening in Indochina. Much to my surprise, it appears that, overall, the DTW framework was found in only 0.3 per cent of cases. In fact, evidence for the framework was present only in *Phase 2* (1.3 per cent), comprising one single instance in the entire dataset, within parliamentary debates.[36]

What can we conclude from this? In the available documents, the DTW framework is clearly absent. Perhaps it predated French unwillingness to carry on the struggle in the 1950s,[37] which increased after a succession of scandals in Indochina was revealed in the press in 1952.[38] However, one compelling reason for why the framework was absent is based on the argument that, in order to have a rallying-around-the-flag effect, the DTW needs to follow from a location that sparks the interest of public opinion. It presupposes that the international issues in question are seen as vital, or at least extremely important, by the public – otherwise there is no diversion to be created.

Poll results suggested that Indochina would have been a very bad choice for any attempt to divert public opinion away from domestic issues. Among all the polls conducted by IFOP, two are particularly interesting and represent a general French uninterest in Indochina. In 1945, IFOP started an investigation into the future of Indochina, a question that had remained unanswered.[39] The aftermath of the Haiphong Incident and the Hanoi uprising in 1946 created ardent discourse in parliamentary debates, with 30 per cent of French people declaring they had no opinion on what was going on. The parliamentary discourses may have displayed a significant importance being placed on the region, although as Alain Ruscio's study of parliamentary debates and polls in France from 1945 to 1954 argues, parliamentary debates could be (very) emotional, but one should not grant too much importance to them because such displays of emotion were normal during the Fourth Republic.[40] Indeed, the emotional tone of parliamentary debates contrasted greatly with poll results. In 1948, IFOP conducted a retrospective study on events that were of significance to France in 1947. The percentage of those who answered "Indochina" was so low that IFOP decided to put them in the category "other" (which together still only comprised 6 per cent of respondents). In this context, it is very unlikely that the various French governments attempted to use Indochina as a diversionary war.

Indeed, further analyzing the IFOP polls, Ruscio shows that the French did not see why they should be concerned about the situation in Indochina: after all, it was a French colony before the Second World War, so it would remain one afterwards. In a newspaper in early September 1945, the socialist MP J. Bourgoin took an interesting and clear position on this: "Was there any thought about the implicit meaning of the expression independence of Indochina? Because Indochina, federated by French arbitration and cemented by French civilisation, would stop to exist the day of its independence."[41] According to Ruscio, 63 per cent of French people thought the same way.

Domestic Legal Arguments Not Related to DTW, But to Avoiding International Involvement

Even though the DTW framework was absent from the dataset, there was still some rhetoric and narratives emphasizing domestic topics. Quantitative analysis reveals that the situation in Indochina was described in 16.2 per cent of cases through domestic legal terms, and in 7.6 per cent of cases through domestic political terms. They were used by all actors, but especially by the government (20.4 per cent and 5.6 per cent), the high commissioner (28.8 per cent and 5.5 per cent), and the military (14.3 per cent and 19.5 per cent).

Qualitatively speaking, what can be said about these descriptions? At the time, it was accepted by almost everyone that what happened in Indochina was part and parcel of French internal – and *not* international – affairs. As such, when the military was sent to Indochina at the end of 1945, they described the situation in terms that recalled domestic debates about the roles of central government and of anarchy.[42]

Events that led to military operations were justified by the need to fight trafficking, particularly in the case of the Haiphong Incident that ultimately resulted in a naval bombing raid by the French navy in November 1946: "Control constitutes only a simple administrative measure aimed to confront the Vietnamese incapacity to stop smuggling … and to protect the piastre by control of [monetary] changes."[43] As Indochina was a colony, if local authorities were unable to tackle issues, it was then the responsibility of the owner of the colony to do so. Any actions conducted against France by the Viet Minh were not described in military terms but rather in domestic legal terms, in the same way that any subversive group would have been.[44]

If the situation was described in domestic terms, then action required to manage the situation was consequently located on the domestic level as well. The military were very clear: the means might be to use the military, but it would work only if there was a clear political plan.[45] Furthermore, French intelligence highlighted the key role of Ho Chi Minh in the deteriorating situation in Indochina. Action was designed specifically to target him and was explained in domestic political terms and not military ones. The point was not to physically eliminate him but to find a local personality who could eclipse him from a political perspective. As General Louis Morlière[46] wrote in a report: "We clearly evaluate that no durable political result would be obtained here if Ho Chi Minh is not neutralised on the Annamite domestic political point of view by a personality enjoying real prestige."[47]

Conclusion

From both quantitative and qualitative analyses of the dataset, it is clear that the DTW cannot explain why France decided to get involved in Indochina from 1945 to 1947. Furthermore, in view of the results, it is also possible to exclude the framework from the ecosystem of motives, reasons, and narratives present on the occasion of this decision. It does not mean that there was no mention at all of diverting French opinion, but simply that it was not found (directly or indirectly) in the available dataset. It is of course possible that there were some such considerations that were simply not captured in the existing documents, or that some documents were not incorporated into the archives and thus disappeared. However, if there were any suspicions that diverting French domestic opinion was a necessity or even of any importance, there would surely have been some early evidence or indication in the dataset. Indeed, the framework was not even found in parliamentary debates, where, if it was a circulating concern, we would expect to have found accusations by MPs that the government was trying to use Indochina to make the public forget about internal problems.

Accordingly, we may conclude that both realist-strategic and DTW-based frameworks of explanation for the French decision to go to war in Indochina do not enjoy empirical support. While some basic elements of these frameworks find minor mentions throughout the conflict, there is virtually no evidence to suggest that they will have had any meaningful impact on the overall decision to go to war. As such, the following chapter will explore the evidence for status and nostalgic virility frameworks to determine whether these competing explanatory frameworks can help us to understand the French motives in the conflict.

4

France in Pursuit of Past Glory in Indochina, 1945–47 (II)

Testing Status-Seeking and Nostalgic Virility Frameworks

This chapter continues the investigation of French motives for going to war in Indochina by testing the frameworks of status and nostalgic virility against the data. A priori evidence for the status and nostalgic virility frameworks was presented in the last chapter. Here, the empirical evidence for these frameworks will be evaluated and placed in conversation with the evidence outlined in chapter 4. We will begin with an evaluation of evidence for the status framework, looking at questions of great power status signalling, followed by an evaluation of evidence for the nostalgic virility framework. By the end of this chapter, we will have a more holistic understanding of the perspectives and motivations leading French decision-makers to engage in war over Indochina.

Intervening to Recover the Glory and Status Lost during the Second World War

Important Status Concerns among the French

Quantitative analysis of French documents revealed that direct and explicit status-seeking ideas and/or concerns were found in 33.3 per cent of cases. First, it appears that the status framework is strongly present in *Phase 1*, with evidence found in a solid 36.8 per cent of documents, but then decreases in *Phases 2* (33.8 per cent) and *3* (29.4 per cent). This suggests that status played a (potentially important) role right from the start, and that it was in France's interest *as such* to engage further in Indochina in any way it could.

Results also show that hardly anyone was wholly indifferent to the idea of status-signalling during the period 1945–47. Evidence for the framework was largely present in documents by heads of state (44.4 per cent) – hardly a surprise, as de Gaulle was leading France's provisional government from June 1944 to January 1946 and made it a priority to restore French status. Notions and ideas surrounding status were also found in a significant number of military documents (44.2 per cent) and in documents from French high commissioners (38.5 per cent), the two most important ones being d'Argenlieu and his replacement, Emile Bollaert. The framework was found at the same frequency in government documents (37 per cent), and less frequently – but still relatively important – in parliamentary debates (22.8 per cent) and the press (26.7 per cent).

The relative importance of status considerations remains to be qualitatively investigated in order to determine whether they were associated with any important decision-making or used as a justification for action. Both such options could be valid given that the framework was present in 38.5 per cent of internal documents and 27.3 per cent of public documents. So, if the question of status was present in many actors' minds, with which narratives was it associated and what role may it have played in the decision to go to war?

French Understanding of Great Power Status: History and Civilization the Characteristics of Great Powers as Seen by the French

There were relatively few direct mentions in the dataset (3.4 per cent of documents overall) of any particular concept of great power status, and there were even fewer explicit mentions of France as a great power (1.5 per cent). Despite being low, these results still provide interesting insights about how great power status was conceived by the French, providing both expected and unexpected indicators.

Expected indicators included the central importance given to the efficiency and strength of France's military (both real and perceived).[1] This was especially pertinent in regard to its expeditionary force sent to Indochina despite post-war turmoil at home: in essence, France was still a great power because it had managed to intervene (in large numbers and far from home) despite a difficult context. From this, the French derived a certain feeling of superiority, as exemplified in a report from general Alphonse Juin dated May 1946: "The influence of the land, aerial and naval forces that constitute our magnificent expeditionary corps goes well beyond Indochina. Countries of the Far East and of our allies have been surprised by this unexpected and significant return of French power."[2] This deployment was qualified a success by some French officials, who claimed that sending the corps enabled France to reclaim its status.[3] In French documentation, one can also clearly identify a feeling of superiority (9.5 per cent overall – increasing throughout the conflict to reach 12.7 per cent at *Phase 3*). Among all the actors, those expressing this feeling the most were the military (11.5 per cent) and high commissioners (15.4 per cent). The feeling of superiority was primarily found in combination with reports of French military victories.[4]

In terms of unexpected results, great power status was, surprisingly, not associated much with the idea of possessing a colonial empire (only evident in 0.6 per cent of documents). Such occurrences were mostly found in the press, which associated a loss of Indochina to a loss of colonial empire and consequential loss of France's position on the international stage.[5]

When analyzing whether the French associated a particular time frame to great power status, the results were fairly uncertain. Among the possibilities *past time frame*, *present time frame*, and *future time frame*, only the first was found (in 3.1 per cent of cases overall). Such discussions were concentrated in parliamentary debates and were made in a rather surprising manner. Instead of clear and direct references to the past, the unsatisfactory nature of the present was highlighted instead. For example, during the parliamentary debates of March 1947, it was noted: "France was absent at the capitulation of Japan. The Americans are the victors of Japan; it is the Chinese army that has occupied Indochina above the 16th parallel and it is the British army that has occupied it in the south. We were no more present in Indochina than we were in Potsdam."[6] Overall, then, we find that great power status, as conceived by the French decision-makers, was surprisingly not constructed around the possession of a colonial empire – which was a key marker of power during the nineteenth century – but more around elements which are key today, such as the ability to project power.

Roles Associated by the French to Great Power Status: Civilizing the World

In addition to their view of what constitutes a great power, the French also had their own ideas about which roles were associated to great power status on the international stage. Despite only sparse quantitative trends, qualitative analysis revealed the perception that a great power's role was to bring civilization to remote parts of the world and to create from scratch a country that met European standards. Such a narrative was not surprising as France's missionary role constituted the bulk of its justification for French colonialism,[7] as well as its justifications for humanitarian interventions in their former colonies.[8] This civilizing role was found mainly in the documents of high commissioners (7.7 per cent) and of parliament (7.9 per cent).[9]

It should be noted that this civilizing role was more evident in public documents (6.1 per cent) than in internal government documents (3.1 per cent). Take for example the 1945 address to the expeditionary corps: "Here, as in any other part of our empire, you will see what has been accomplished ... Numerous schools have been opened and the benefits of our civilisation have been brought to the most remote regions."[10] In the few mentions of the civilizing role in internal documents, the idea was linked to the conduct of military operations, especially to the protection of minorities from the Viet Minh.[11]

Between Displaying Military Strength and the Need Not to Antagonize Key Allies: Divisions over Which Appropriate Status Signals to Send

Several codes were used in the analysis to measure whether status considerations were seen as important with regard to the situation in Indochina. The first, *situation understood in reputation/status terms*, was present in 13.1 per cent of cases overall (17.9 per cent in *Phase 1*, decreasing to 10.4 per cent in *Phase 2*, and 9.7 per cent in *Phase 3*). The *need to send appropriate signals* was found in 11.9 per cent of cases overall with a strong presence in *Phase 2* (15.5 per cent) compared to *Phases 1* (12.1 per cent) and *3* (8.8 per cent). The last code used was *location seen as important for status*, which tagged any explicit association of the word "Indochina" to any status-related idea. It was found in 9.1 per cent of cases in *Phase 2*, before decreasing to 3.9 per cent in *Phase 3*.

Overall these results highlighted that status considerations were important *before* the situation worsened. They suggest that, based on the decrease over time of status concerns, France's involvement in Indochina had the objective of addressing these status concerns (the frequency of which decreased once there was active involvement).

Qualitatively, these status-related codes were associated with multiple narratives. The first narrative was that the intervention in Indochina was meant to atone for Japanese humiliation and reinforce France's claim to be one of the victors of the Second World War. The second was that France felt anything but certain of being supported in its attempt to regain control over Indochina and therefore had to send specific signals either to gain support or to decrease hostility. The final narrative was that, from a French perspective, the situation was extremely complex because other actors also had reputational interests at stake in Indochina. The following sections will outline each narrative in turn.

The Temptation to Send a Strong Military Signal

The idea that Indochina was of importance in terms of France's reputation and status is of significance for various reasons. First, because of the damage caused by Japan in its attempt to get rid of all French resistance in Indochina before it eventually capitulated. Every French garrison was attacked by Japan on 5 March 1945 and the survivors were executed.[12] The military reported that French status had been destroyed on that day: "The prestige of the white man has been seriously touched by the Japanese and by the occupation … Absent in Indochina at the moment of Japanese capitulation by the will of incomprehension of our allies, at this essential moment we have not appeared as victors."[13]

From this moment on, every possible aspect of French status was at risk. The timing of public declarations by French officials and the (mis)use of words could be interpreted as another display of weakness. This was the case especially in *Phase 1*, when it was discussed whether France should make a second declaration regarding the future of Indochina, just after a first one declaring that some autonomy would be granted to Vietnam within the French-dominated structure of a *Union Française*. The declaration carefully avoided any mention of the word "independence" or of any situation that might imply it.[14] In his instructions, de Gaulle ruled out any new declaration: "The fact is that to make a new one in the current circumstances would have all the appearance of a retreat, which would make the situation in and outside Indochina yet more complicated."[15]

Strategic and tactical military decisions were also analyzed in this light, the most noticeable example being found in the report by Colonel Pierre Louis Debès of the events in Haiphong,[16] where he argues that the agreement reached goes against French interests: "French and Vietnamese troops will return to their respective positions. This second condition is in fact equivalent to permitting the Vietnamese to re-occupy all the positions from which they have been expelled."[17] Two days later, a report issued by Morlière warned the French government that "only a determined attitude by the French government and its decision to accept and support a prolonged and serious effort can lead to saving French Indochina."[18] Adopting a firm military posture was seen as beneficial to gaining local public support, especially in areas known to be hostile to the Viet Minh.[19]

Showing Composure to Gain the Support of Allies, or at Least Avoid Their Increased Hostility

The situation might have seemed simple: flex military muscle and all will go back to normal. But this ignored an important part of the broader picture: the hostility from other countries. France was torn between the hard-military line required to gain superiority in Indochina and the need to send different signals to calm unfriendly international opinion – a problematic discussed in 14.1 per cent of cases overall. Explicit mention that France was not supported by other countries was found in 9.2 per cent of cases. Additionally, in 12.5 per cent of cases two countries were specifically identified as villains: China and the US.

The literature emphasizes the US's hostility to a French presence in Indochina. One can find such mention in the dataset as early as 1945 when Office of Strategic Services (OSS) officers, deployed in theory to support American troops, were found to have received very different and anti-French orders.[20] Specifically, it was President Franklin Roosevelt who was most antagonistic to a French presence, and the French seemed to have understood that.[21]

The opposition of US units to French troops on the ground was so evident that it culminated in official diplomatic protests through various channels, such as a telegram sent by Leclerc in September 1945: "I strongly urge you to protest in Washington against the following facts: Primo: General Gallecher, the US representative to General Lou Han [the Chinese general in charge of the occupation of the northern part of Indochina] in Hanoi has declared that French sovereignty over Indochina was not recognised by the Allies and that it was out of question that a French protectorate in Indochina should be restored."[22] Ultimately, Roosevelt's death on 12 April 1945 proved to be very welcome to the French.[23] An improvement in relations can be seen quantitatively in the dataset, where the code *other country as villain* was found in

22.2 per cent of documents in *Phase 1* and then decreased to 9.1 per cent in *Phase 2*, leaving China as the country most opposed to a French presence in Indochina.

In an attempt to decrease US hostility, or at the very least not to increase it, the French, especially the military, were conscious that they had to send targeted and appropriate signals. From 1945, they sent instructions so that the re-establishment of their administration in Indochina would go as smoothly as possible: "It is essential to avoid any problems with this establishment that could be exploited against us on the international stage."[24] The particular attention France gave to noting how international opinion reacted to events in Indochina is a leitmotif, with much evidence being found throughout 1946 and 1947.

The worst-case scenario identified by the French – and particularly mentioned in 1947 – would be for the situation in Indochina to be discussed at the new organization, the United Nations.[25] On the last day of his premiership, President of the Council[26] Paul Ramadier stressed during an interdepartmental committee that long-lasting military operations could have negative repercussions on international opinion, especially American.[27]

Any military campaign would have to be short and decisive. The objective was to keep the dispute strictly at the domestic level to avoid interference from other countries. It was felt this could only be achieved through a restrained use of military force. French soldiers were warned: "It is our duty, via our conduct, to refuse any pretext to our adversary that might worsen the situation and might justify the aggression policy of the Viet Minh. One of the most effective themes that the Viet Minh is using with the French … and at the UN, is the extreme brutality with which, according to them, we would be conducting military operations."[28]

Apparently, such caution was not enough: MPs became impassioned and started to discuss the *New York Times* editorials written on the situation in Indochina.[29]

Intertwined and Opposing Symbolic Considerations: The French Were Not Alone in Having Status at Stake in Indochina

The initial situation became more complex and harder to manage due to France's opponents involved in Indochina, namely the Republic of China and the Viet Minh, who also had symbolic interests at stake. When France tried to re-establish its position in Indochina, it had to deal with the fact that the colony was under British administration under the 16th parallel and under Chinese administration above the 16th parallel. In addition, there was the relatively precarious internal situation in which Chiang Kai-shek[30] was always trying to establish full control over the military.[31]

In the eyes of the French, Chiang Kai-shek was following a status-related policy in Indochina and seeking quick success to strengthen his position against the communists.[32] The presence of Chinese troops proved also to be a challenge for France, especially when it tried to regain control over northern Indochina.[33] To achieve their objective, the French military framed the situation in symbolic and reputational terms: "Avoid vexing them but instead flatter their vanity … We never insist enough on the importance to flatter Asians no matter if they are Chinese or from the Viet Minh."[34] The presence of Chinese troops also created troubles for the French deployment and their presence weakened France's reputation: the longer Chinese troops occupied parts of Indochina, the longer France would appear weak and incompetent.[35]

On top of dealing with Chinese pride and its quest for status, the prestige, status, and reputation of the Viet Minh were also at stake – already identified in 1945 as a major difficulty to achieving a solution to the dispute.[36] At that time, one question for Viet Minh officials was how to capitalize on the independence granted by the Japanese just before their retreat and to make it acceptable to the French.

During direct negotiations in Indochina with Ho Chi Minh, France (through General Raoult Salan) tried to play the prestige and reputation card in an attempt to flatter and make him accept a compromise.[37] This approach was not successful as expressed in Morlière's report about the Haiphong Incident in December 1946: "We are at a dead end. The government of Hanoi seems to be unwilling to accept, without a new test of force or counterparts – but which ones? – the conditions it has been offered in Haiphong … To accept this loss of face … it is condemning itself to fall apart and take in its fall the entire Viet Minh party."[38]

Conclusion

Quantitative and qualitative analyses of the dataset revealed a strong status-related component to the conflict and for France's decision-making therein. It would appear that the Second World War, and most importantly the consequences of the Japanese coup in 1945, played a key role in damaging – one might even say destroying – what was left of France's great power status in Asia. French officials assessed the situation based on the consequences of the Japanese coup and policies were defined with the objective of restoring French prestige based on a redefined civilizing role.

Every move was either understood or defined by status and prestige considerations – from military operations and tactics to negotiation strategies. Furthermore, it appears that the situation was rendered more difficult than planned because France's adversaries and opponents were also engaged in the pursuit of status in Indochina. Confronted with a sort of symbolic zero-sum game, it appears that none of the actors could give up their objectives without suffering very damaging and unacceptable symbolic consequences. It would seem fair, therefore, to claim that status considerations were not only important, but played a key role in the French decision to go to war.

Nostalgic Virility in Indochina: A Fear of Castration?

Strong Empirical Evidence

Taking into account codes measuring both the immediate/direct aspects of the framework and those measuring indirect aspects, evidence for the nostalgic virility framework was found in some 55.4 per cent of documents. These results make the framework the most (quantitatively) dominant in the dataset among those tested. Even if we take into account only the codes directly measuring specific aspects of the framework, the results demonstrate a still strong 47.7 per cent presence across documents.

The nostalgic virility framework is more prominent in *Phases 1* (46.5 per cent) and 2 (57.1 per cent) compared to *Phase 3* (45.1 per cent). This could be explained by historical events that happened during the first two phases: the Japanese coup, the Haiphong Incident, and the Viet Minh attack on Hanoi. Once the decision to go to war was fully acknowledged, the French elite could feel "reassured" about their virility, hence a decrease in the frequency of such decision-making logics.

Who seems to be most concerned by nostalgic virility considerations? The framework has the highest frequency in military documents (63.9 per cent), followed by those of high commissioners (59.6 per cent) and heads of government (55.6 per cent). Perhaps the biggest surprise

Table 4.1 Frequency of the four theoretical frameworks in the Indochina case

Framework	*Frequency (per cent)*
Material and strategic	2.8
DTW	0.3
Status and symbolic	33.3
Nostalgic virility	47.7

here is the high frequency in the latter. From 1945 to 1947, France had five heads of government from three different political parties. First was Charles de Gaulle, and it was no surprise to see him linked to the idea of virility. But this remained strong with Félix Gouin (socialist), George Bidault (Christian democrat), Léon Blum, and Paul Ramadier (both socialists). It appears that political differences had little impact on the use of this form of reasoning.

A Highly Emotional Situation Closely Associated to the State of French Military Forces

Quantitative analysis of French documents revealed that emotions were evident in 26.9 per cent of cases overall (most importantly in *Phases 1* and 2 with 25.7 per cent and 33.8 per cent, respectively). The emotions at play range from feelings of weakness and loneliness to disgust and betrayal, with occasional signs of pride. A key characteristic of the emotions expressed is their association to the state of the French military, many of which appear to be the consequence of its position in Indochina.

Another key characteristic of these emotions is that they were associated to other issues (to be addressed in later sections) such as nostalgia and the desire to punish morally deviant behaviours. No main group of actors escaped the effects of these emotions: heads of government (22.2 per cent), members of government (18.5 per cent), high commissioners (26.9 per cent), the military (27.3 per cent), and the public (MPs and the press, 31.3 per cent) all showed strong evidence of emotional affect and reasoning.

Initially after the Second World War, the French felt a mixture of weakness and loneliness linked to the position of its army. The literature is in agreement that the delivery of shipping equipment and rations to the

French was not usually a priority of the Allies – and that when it was, the goods were often pilfered by China – thus showing that the French government was not one worth taking seriously at this time.[39] Despite this precarious situation, there are a few mentions of feelings of pride associated with the rare French military successes against the Japanese.[40]

Feelings of pride temporarily increased when the French army's motivation remained high even in difficult conditions.[41] The feeling of loneliness, or at least of not being supported by the Allies, was also found in parliamentary and press discussions of the situation. MPs complained that the colonial system was not being defended in front of critics, and this despite the massacres committed by the Viet Minh: "At a moment when campaigns in France and abroad have not stopped the attack on 'French colonialism,' a yellow book about the atrocities committed by Vietnam on our compatriots in Indochina has not yet been published."[42]

Feelings of humiliation were exacerbated through the situation created by the Japanese forces, and this just before their capitulation.[43] Even worse, after the capitulation, Japanese forces could keep their weapons and were tasked by the Allies to police Indochina, while French soldiers, who were then freed, remained unarmed and stationed inside their garrisons.[44] The landing of the French expeditionary corps at the end of 1945 did not resolve the feeling of humiliation felt by the French living in Indochina, but in fact increased it. The profound difference between them and those landing came from the French political divide during the war. Indochina, like most of the empire, remained under control of the Vichy, and was thus following, more or less, a path of cooperation with Nazi Germany and Japan. The expeditionary corps was essentially composed of units that rallied around de Gaulle in London after 18 June 1940. The expeditionary corps went to Indochina with a feeling of superiority based on their participation in the victory against Germany and associated the former to the losing side because of Vichy collaboration.[45] The division was also captured in the French documents (especially after de Gaulle's departure): "The French of Indochina keep a very bad memory of the French landing, because they have been hurt by the attitude, often unjust and hostile, of compatriots who they had awaited a very long time."[46]

As seen above, emotions were diverse and found in many contexts. One emotion in particular seems to have stood out and effectively summarizes the overall French behaviour: the humiliation which gave rise to a need to redeem the name and honour of France. This emotional response resulted in efforts to show other countries that the French military was competent and skilled and to the locals that the "real French" – meaning those who did not surrender – would arrive to take charge.

The Central Place of Virility in Thoughts and Actions

Qualitative analysis reveals that virility is present within two opposite narratives. The first narrative is positively constructed and attempts to show "what" French virility is. The second is negatively constructed and highlights direct and indirect concerns "about" virility.

Virility becomes an important issue first and foremost when French virility is questioned and challenged. The issue of appearing strong was found in 10.7 per cent of cases and remained quite stable throughout each phase of the conflict. Fear of symbolic castration comes from the fear of losing control over objectified others (8.3 per cent), which itself comes from a feeling of having a weakened military (code: *military as victim*, 9.2 per cent of cases). The powerlessness of the military was symbolized by the need to efficiently protect civilians, which they were not doing (code: *population as victim*, 9.2 per cent of cases). The cause of this powerlessness was detailed to George Pompidou at the end of 1945, highlighting that if the French were able to resist the pressures of the Viet Minh it was because of the covert use of British troops: "We are reduced to powerlessness by the lack of modern weapons, of ships, of planes and of troops. We could not have resisted alone against the Viet Minh Army. We use the British army as a protecting screen without telling the Labour government."[47] It was unsurprising, therefore, to find a relatively important unconscious virile discourse in French documentation (19.3 per cent overall), which increased as the situation in Indochina developed. All of this shows that the French not only referred explicitly to ideas of virility, but also that they used a virility-related lexical field in general. They used an explicit virile narrative centred around the display of traditional characteristics of virility (especially in military documents: 18 per cent). This narrative is constructed around the ideas of *strength* (11.9 per cent) and *being firm and resolute* (8.3 per cent). *Being brave* is also present but is rather peripheral (1.5 per cent). The other characteristic of this positive virile narrative is that it had a dual construction: on one side the idea of strength, mostly found in military documents (24.7 per cent), and on the other the idea of firmness and resolve, essentially found in documents issued by politicians (9.2 per cent when combining the variables *member of government, high commissioner, MPs*).

So, the narrative is constructed around the idea that although the military had to fight in difficult situations, they did have some success and their opponents could not do much about it. Military strength was emphasized in many different ways. According to Valluy, the simple fact of seeing French troops made the Chinese panic and feel much less confident.[48]

At perhaps the tensest moment in 1946 (when the French were besieged in Haiphong), French troops were described as not particularly afraid or psychologically affected by the threat: "During these two days, all units of the garrison were magnificent, running into the battle singing, looking down upon death."[49] Such a display of strength and virility was also present in *public documents* (14.9 per cent), especially in letters from generals. One from Valluy is particularly and explicitly virile: "What is our goal? To strike at the top! We should have done this in May had it not been for tiredness, the heat, and the turnover of troops!! It is important to fix [to be understood in military terms] the Việt Minh government, to catch it if possible, and if not to pursue it tirelessly … until it surrenders."[50]

Newspapers, while possibly in more neutral terms, also highlighted the superiority of the French military, such as in a *Le Monde* editorial of January 1947: "Elements of enemy troops tried without success to infiltrate a French outpost."[51] As a result, the military situation appeared to the French as positive and under control at the end of 1947. French offensives were qualified as "powerful" in reports to the government,[52] whereas in the conclusion of his report (and although he points out the risk of ambushes and skirmishes) Morlière noted: "Militarily, there is no massive fear of failure in Indochina. Our superiority has been affirmed right after 19 December and the important forces that we have cannot be disallowed. We will go where we want, when we want."[53]

The second traditional characteristic of virility is to be firm and resolute. In the current context, this meant France refusing to give up Indochina and sticking to an established strategy (found in 8.3 per cent of cases overall). In 1945, the situation did not look good, although two years later, in 1947, the tone of French comments had changed radically. Improvements in the situation with regard to a more favourable view of

France were explicitly associated with the strength of the military and also the firmness and resolve of all the actors engaged. Representative of this is the open letter to reinforcements written by Valluy in February 1947, although doubt could be cast as to the full authenticity of any document addressed specifically to the public. Does the document really describe exactly what was going on when its objective was to motivate the troops? Given the general tone of all the other documents, it is likely, as it specifically identified resolve as the key factor that enabled France to improve its situation in Indochina: "One year ago, France saw her intervention in Indochina encumbered by a double foreign mortgage … At any moment the action taken to re-establish its position risked being judged, controlled, even paralyzed by the Allies responsible for peace in southeast Asia. She has recovered right at the end of the first semester because of her patience of 1946 her freedom of action."[54]

Politicians regularly expressed their will to maintain the French presence in Indochina. Bidault, then *de facto* head of government, emphasized this determination: "We have to be firm in indicating that France will not leave Indochina and that she will defend her presence with all the means available."[55] MPs were not to be outdone on the matter. The 18 March 1947 parliamentary session was very representative of how emotional parliamentary interventions in regards to foreign policy and colonial decisions could be – despite their relative unimportance in terms of influencing the government. Affirming that "France must stay in Indochina," it was reported in transcripts that all MPs applauded, whether from the right wing, left wing, or the centre.[56] In later debates during the same parliamentary session, some MPs argued that independence was to be decided and granted by Paris alone and not by the Indigenous people.[57] Such a show of resolve underlines real virility-related concerns.

Tensions about Understanding Virility: Punishing or Nurturing?

This section draws heavily on the work of George Lakoff and attempts to operationalize his thoughts into the field of IR. He argues that our political world views are structured around two implicit sets of metaphor: the strict father type and the nurturant parent type.[58] How would these metaphors be operationalized in the specific context of French involvement in Indochina?

Table 4.2 Distribution of the narratives based on the variable "actors"

Actor	*Strict father narrative (per cent)*	*Nurturant parent narrative (per cent)*
The military	23.4	5.2
Members of government	20.4	9.3
High commissioner	34.6	11.5
MP	14.9	7.9

Due to the context and nature of the conflict (a defeated European colonial power trying to re-establish its position in a colony), one would expect to find the two mindsets identified by Lakoff: a strict-father-oriented discourse arguing that the Viet Minh should be punished for their actions, and a nurturant-parent-oriented narrative probably associated to the idea that Indochina needed help from its colonial power to achieve its full potential.

Quantitative analysis reveals that these two sets of metaphors are present in the dataset, although the strict father narrative is more frequently identified (21.4 per cent) than the nurturant parent one (8.9 per cent). Furthermore, it appears that the most significant actors use both narratives, as the table below summarizes.

It seems that all actors agreed on the core of each narrative: to punish morally wrong and deviant behaviours and to protect the weak from "evil" (i.e., the strict father narrative), and to help Indochina grow through a positive power relationship (i.e., the nurturant parent narrative). However, some differences were observed when it came to who exactly should be punished, as there was an external/internal divide structured around different hero-villain-victim narratives.

A Strict Father Perspective: A "Downgraded" and "Fake" Virility to Be Punished by a "True" Virility

The strict father narrative was identified in 21.4 per cent of cases overall, especially in *Phases 1* and *2* of the conflict (22.9 per cent and 24.7 per cent, respectively). It was essentially built around three main ideas: (1) the self-descriptor of an actor being mature and disciplined, which

(2) gave France the authority to punish deviant behaviours and (3) protect the weak from a political "evil." These ideas divided and separated the French from who they saw as their deviant and barbaric opponents, namely the Japanese, the Chinese, and the Viet Minh.

From the dataset, it seems that the French got their legitimacy to punish deviant behaviours in Indochina precisely because they behaved differently: they were mature and disciplined – that is to say, they scrupulously abided by laws and rules of engagement on the one hand while doing their best to avoid unnecessary casualties, especially military, on the other. Soldiers in the expeditionary corps were briefed on this: "Do not forget that any looting is forbidden and will be dealt with the gravest sanctions. No unnecessary destruction, no brutality against nonfighters, women or children."[59]

The content of self-description evolved over time. Initially, it was based on the capacity of the military to remain calm and disciplined whatever the situation, even in front of very aggressive Chinese and Viet Minh provocation[60] – the latter of whom targeted them randomly with artillery in attempts to make them seek revenge, which would result in civilian casualties.

Being mature and disciplined also meant following protocol when dealing with opponents. When it was still a possibility that there would be negotiations with Ho Chi Minh, the French treated him based on his rank as President of Vietnam: "The gesture of Admiral d'Argenlieu welcoming with all the required solemnity and cordiality President Ho Chi Minh on his way back to Tonkin, constituted a first response to this campaign in showing that we never hesitate to make the first moves on the path to understanding."[61] Finally, when it was obvious that the Viet Minh were not playing by the rules and were not enforcing the negotiated agreement, France did not use this as an excuse but instead continued to carry out its part of the agreement: "France will never disown any of her promises under the lame excuse that those who signed the agreements happened to be unworthy and failing."[62]

On the other side were those opposed to France. They are clearly identified (apart from the UK and US at the very beginning) from analysis of the dataset. Two codes appear most frequently: *opponent leaders* (used to tag any direct or indirect mention of Ho Chi Minh and the Viet Minh: 19.9 per cent) and *other country* (used to tag any reference to any country, apart from Indochina, that were perceived as an adversary: 12.5 per cent).

Qualitative analysis reveals that the other countries in question were essentially Japan and China. The common point between Ho Chi Minh, the Viet Minh, the Japanese, and the Chinese was that they were all described as "faking" virility: having and demonstrating either personality traits and/or behaviours in opposition to or negatively mirroring France's expectations of virility. The code *opponent seen as castrated/ 'fake' male* was found in 20.8 per cent of cases overall. How did the Japanese, the Chinese, and the Viet Minh come to be qualified as such?

After Japan had surrendered, it was tasked with policing Indochina until the arrival of French troops. Instead, they appeared to continue the war informally: "They sneakily continued the fight they should have officially stopped."[63] They were then accused of trying to destabilize France as much as possible by liberating criminals, including members of the Viet Minh who had been imprisoned during the occupation of Indochina.[64] The Chinese, especially local officials, were regarded as acting duplicitously and seeing the occupation of northern Indochina as the perfect opportunity to increase their personal wealth.[65]

They were associated with the idea of downgraded virility for two main reasons. First, they (Chinese leaders and officials, especially Chiang Kai-shek) did not meet the traditional characteristics of virility: strength, courage, and firmness. Symbolic of this was their incapacity to be obeyed by their own troops, which impacted their ability to negotiate and left the French unable to take anything they said for granted.[66] The Chinese were also accused of being duplicitous: speaking in favour of France while at the same time helping and supporting Indochinese groups hostile to France by supplying them with weapons.[67]

Ho Chi Minh and his Viet Minh Party faced the most criticism, and from all the French actors analyzed in this study: heads of government, ministers, the military, and MPs. They were accused of being anything *but* brave and courageous, and of lacking courage on the battlefield, as can be seen in the report from Colonel Debès about events in Haiphong: "Because of our insistence, a Vietnamese officer, Lieutenant Trung, from the liaison, is finally sent to the field to tell the defenders to cease fire. He hides in a cave from which he emerges only the day after at 10 o'clock."[68] Viet Minh officials rejected any responsibility for occurrences in Haiphong and rejected that the local population had attacked French convoys.[69]

It was not because the Viet Minh had independence as a political objective that it had the support of, and wanted to help, the local population – quite the opposite, according to French documents. In the dataset, the Viet Minh were often described as untrustworthy. From a humanitarian perspective, reports suggested that they abandoned the weak and injured on the battlefield (a clear violation of the Geneva Conventions). According to military reports, when clearing the city of Haiphong, the French troops found the situation shocking. Debès's report is clear and explicit: "The search in Vietnamese quarters ... has enabled us to penetrate inside the Vietnamese hospital. It was deserted right after 24 November by Vietnamese doctors and medical staff. The rooms offered an appalling spectacle of injured, sick, dead bodies already in putrefaction, all intermingled."[70] From a political perspective, at first the Viet Minh were allied to other groups and movements, but this did not last long as they had a tendency to "use" them to cover their military retreats. Valluy noted "The contempt with which, during the recent Tonkin campaign, the Viet Minh sacrificed the Tu-Ve and local militias in rear-guard action to assure the retreat of the 'pure.'"[71] Because of this, and throughout the conflict, members of the Viet Minh were described as "fanatical,"[72] "terrorists,"[73] and "criminals."[74]

Another aspect of what was considered downgraded virility was the permanent change of the Viet Minh's political decisions, which contrasts in the dataset with the French self-descriptor of being mature and disciplined by following rules and keeping one's word. The Viet Minh were accused of engaging in negotiations only to gain time to strengthen its military position and confuse the French about its real intentions: "We find, next to the obvious façade of a conciliatory and pacific attitude, a constant will to continue the fight to achieve well-defined objectives."[75] The slippery attitude of Ho Chi Minh was reported by historians as seriously annoying the French government, to such an extent that in August 1946 Bidault was advised by Pierre Messmer, secretary of the inter-ministerial committee, that "it was time to put Ho Chi Minh in front of his responsibilities."[76]

The tone of the French media and in parliamentary debates was even more explicit. In parliament, MPs made use of the Second-World-War analogy of Pearl Harbor to describe and characterize the situation. Right before the attack, Japanese officials were telling their US counterparts that Japan was a pacific country that only wanted peace. According to French MPs, France was the target this time.[77] So, the feeling in France was that, because of their lack of humanity and empathy as well as for their treachery, the Viet Minh had to be punished (severely), and the population protected from them.

Quite surprisingly, these two ideas were found less frequently than expected in the dataset. Punishing deviant behaviour was found in 6.4 per cent of cases and the idea of protecting against an evil actor in only 4.6 per cent of cases across the dataset. Despite the low frequencies, a clear change in French posture was observed both quantitatively and qualitatively. In *Phase 1*, giving protection against an evil dominated the strict father narrative (6.9 per cent), while the idea of punishing deviant behaviour was secondary (4.9 per cent). *Phase 2* witnessed a reverse in priorities, with punishing deviant behaviour becoming primary (9.1 per cent) and the idea of giving protection against an evil secondary (3.9 per cent).

This change of approach is also noticeable qualitatively. Reports in the dataset argue that a conciliatory approach was not enough and that a more direct and expeditive one was required. A first inflexion in the approach can be found in the report justifying the intervention of French troops in Haiphong. The main motivation was to protect civilians, but with the implicit idea of putting an end to what threatened them.[78]

In February 1946, Ho Chi Minh had been personally warned by Salan that the threat to civilians constituted a red line.[79] In Valluy's report of the situation at the beginning of December, the switch from protecting to punishing was complete as he rejected the conciliatory strategy of Morlière, de facto opening the door to a more direct approach.[80]

Political and Virile Divisions through the Strict Father Narrative: The French Were Not as United as It Seemed

Qualitative analysis revealed that the strict father narrative was also associated to a hero–villain narrative at the domestic level, capturing operational and political divisions.

As evidenced in the literature, there were tensions between the government and d'Argenlieu about how things should be conducted in Indochina.[81] This villain narrative was found only in *Phase 1* (10 per cent), corresponding to the period of time when d'Argenlieu was the French high commissioner. He was accused by the government of conducting his own policy in Indochina without asking for approval or informing Paris of his actions.

On 13 March 1946, he was officially reprimanded for this after he made it clear that he rejected the agreement Paris had negotiated with Ho Chi Minh. D'Argenlieu was reminded that there was only one policy for Indochina, and that was the one defined in Paris: "The government is now in front of a *fait accompli* regarding issues it was never consulted about. In the future, no agreement, regardless of its nature, can be made without the text being approved by government."[82]

The government was also portrayed as villain by some MPs (11.9 per cent), almost as often as the identification of Ho Chi Minh and the Viet Minh Party as villains (14.9 per cent). Parliamentary debates showed that the press (5.9 per cent) and other MPs (6.9 per cent) were identified as villains too. These narratives reveal the existence of political tensions over the situation in Indochina: they were used to criticize the actions of the government, the press, and of other MPs towards the military.

MPs and the government tended to identify the press as villains in a similar way. The press was accused of being partial and of giving what would nowadays be called "fake news."[83] Such misinformation was clearly identified as a major reason why France was struggling to find an agreement in Indochina: "I have to stress how frequently – because of the exaggerations by the French press and press agencies – the populations of Indochina … have more and more the impression that France will link their entire future to negotiations with the Vietnamese government alone."[84] Later (*Phases* 2 and 3), only MPs criticized the press. Parliamentary critics were stronger than those of the government as they accused the press of high treason by providing wrong information.[85]

Criticism got more virulent – of the press and the government – after some newspapers published a manifesto from a trade union describing the French military as Nazis and implying that the government was doing nothing to prevent its behaviour.[86] However, the strongest criticism of the government was linked to the way, according to MPs, it treated the military. It was accused of sacrificing the lives of soldiers engaged in Indochina for nothing, after some newspapers reported Blum, the new head of government, saying that the only solution was to negotiate. Consider this declaration of 12 December, after the Haiphong Incident but before the general attack of the Viet Minh (19 December): "Two days ago, Mr Léon Blum, in a first-page article wrote: 'in my opinion, the only solution is a sincere agreement on the basis of independence.' Mister Minister of French overseas territories, tell this to the 80,000 soldiers who are fighting there."[87]

The government was also accused of playing tricks on the military engaged in Indochina by submitting them to high postal and customs fees when they corresponded with their families in France.[88] Some MPs were also accused of duplicity by their colleagues: saying publicly during parliamentary debates that they were supporting the military but in practice refusing to vote for military credits.[89]

Finally, the analysis of the uses of the strict father narrative captured an opposition within the French military: both d'Argenlieu and Valluy opposed Morlière. D'Argenlieu and Valluy were well-known for favouring an "energetic approach," although here Valluy preferred a more cautious one. Indeed, it seems that the differing views about strategy revealed an implicit difference of opinion about the notion and definition of virility.[90] On the one side were those who showed signs of nostalgic virility, namely those who wanted to punish the Viet Minh hard and fast. On the other side were those who acknowledged that behaviours and approaches to the situation in Indochina had to change, and that what was once permitted was no longer appropriate in a post-Second World War context – implying that ideas about virility and virile behaviours also had to change.

In his reports to the prime minister, the ministers of defense and overseas territories and to general Juin, D'argenlieu strongly criticized Morlière. D'Argenlieu used virile terms to criticize Morlière's cautious approach and to accuse him of passivity. D'Argenlieu then continued by arguing that Morlière's approach was the reason why the Viet Minh were growing stronger: "I have directly intervened to General Morlière to incite him to adopt as response to the Government of Hanoi's provocations a more energetic attitude and not to encourage by his passivity the extension of works that threaten more openly day after day the security of the population and of our troops."[91]

After d'Argenlieu was removed from office on 5 March 1947, opposition between nostalgic virility and "modernist" virility continued through Valluy and Morlière. Valluy had always been in favour of taking impeditive measures against the Viet Minh, while Morlière appeared to favour a more cautious approach. Valluy's preference for an energetic response to Viet Minh actions is well known, as was his push for authorization by the French government.[92] The Valluy/Morlière opposition has been well documented by historians: Valluy even bypassed Morlière's chain of command and gave an order directly to his subordinate, Debès, to engage Viet Minh units during the Haiphong Incident. Debès was one of those more willing to engage strongly with the Viet Minh and obeyed with delight.[93] The instructions Valluy sent were very explicit: "The moment has come to give a harsh lesson to those who traitorously attacked us."[94] Historians have shown that the military in Saigon, organized around the figure of Valluy, constituted a "bellicose clan" who shaped the French government's opinions and perceptions of the events through controlling what information reached them. As a result, the contradictory report from Morlière (see below) reached Paris long after the incident.[95]

Morlière's approach is completely different. In his report, he returned to the Haiphong Incident and, although Valluy is not mentioned, the criticism is explicit: some French in positions of power live in the past and their decisions are counter-productive at the very best: "Quite unimportant in its origin, this incident, at first de-escalated with great difficulty, escalated again due to a tense local atmosphere because of circumstances and persons and also due to direct and inappropriate and dangerous intervention by the Saigon authorities ... And because of some French civilians who are stuck in the past and will never understand that some epochs have gone for good."[96] Based on this report, and given his opposition to Valluy, who clearly qualifies as a nostalgic virile individual, it would be tempting to see Morlière as a sort of "modernist" virile individual. Building up the opposition between the two views taken on virility might serve the overall nostalgic virile argument, but things are never that easy or clear. In a second report, Morlière also pleads for the French government to show a more determined attitude.[97]

If labelling Morlière with modernist virility might be too much given his wish to safeguard the French colonial empire, it is nevertheless possible to say that he seems to offer a different understanding of nostalgic virility to the one of Valluy. The historian Philippe Franchini noted that part of the hostility to Morlière was due to his being more of an office man,[98] which could explain his different and more cautious approach during the Haiphong Incident. It did not, however, prevent him from illustrious action much later during the campaign and to being cited six times to the *ordre de l'Armée*.

A Nurturant Parent Perspective: Helping Indochina to Grow on the Path to Adulthood

According to Lakoff, the second set of metaphors structuring political discourse is the nurturant parent narrative, identified here in 8.9 per cent of the dataset. This is constructed around the idea of developing a positive power relationship between France and the different entities composing the colony of Indochina, something which all French actors (government, military, and parliament) agreed was necessary.

There was no real evolution in the qualitative construction of the narrative; it was more so a traditional colonial narrative whereby the colony needs its metropolis to achieve its full potential, as suggested in 1945 by the high commissioner: "It is a question of reconstructing true French protection for the Indochinese populations, that is to say to help them mature politically."[99] Even more explicit is the formulation found in reports of the situation in Indochina in July 1945: "France engaged many times to protect and guide them to a civilization that would bring a better quality of life."[100]

However, there was a certain amount of uncertainty regarding this narrative: was it genuine or only for purposes of propaganda? In *Phase 1*, when France was trying to re-establish its position in Indochina (which meant getting the approval of allies as well as some local domestic support), the frequency of the narrative was at 9 per cent. In *Phase 2*, when France was deciding which position to adopt to respond to the Viet Minh attacks in Hanoi, the frequency was similarly at 9.1 per cent. But in *Phase 3*, after it became clear that no agreement would be possible and that the Viet Minh were not interested in such an agreement – and that the only response now would be war – the narrative frequency dropped to 7.8 per cent of cases. An explanation of this decrease could have been that in *Phase 1* the narrative was used to gain external support and decrease potential local hostility to a re-establishment of French legitimacy. In *Phase 2*, the narrative was used to gain local support in the upcoming fight against the Viet Minh. Finally, in *Phase 3* it was used in the same way as *Phase 2*, but was now less necessary (and perhaps overshadowed by strict father narratives).

Finally, we see that this narrative was used in different ways based on the type of documents analyzed. In internal government documents, the frequency was 6.3 per cent, whereas in public documents it was 9.7 per cent. The content was also impacted by the type of document, not from the perspective of the underlying ideas but rather in the tone used (usually more grandiloquent and emotional). For example, Bollaert (who replaced d'Argenlieu as high commissioner for Indochina on 13 March 1947) said in a speech: "France offers the populations who previously made up the French empire the opportunity to move in an appropriate environment alongside France, as equals, inside a majestic organization which will enable them to grow while being faithful to their own spirit."[101] Another representative formula can be found in the report sent to the government by the new high commissioner, Emile Bollaert, in which he advises: "The era of imperialism is over, the era of friendship begins."[102]

The presence of a nurturant parent narrative captures the evolution at this time in the conception of colonialism. The international context – the power transition in favour of the US and of the USSR – made it very difficult for European powers to maintain their rule over colonies and the idea of more balanced relations between the metropoles and the colonies started to emerge as an attempt to safeguard interests in the colonies.

Conclusion: What to Think about the Nostalgic Virile Framework

From the quantitative frequencies of nostalgic virile arguments and the qualitative analysis of the dataset, it was possible to affirm that the nostalgic virile framework was an important analytical tool and helps provide a strong explanation for the French decision to go to war in Indochina. Several key aspects of the framework were present, primarily the ideas of appearing strong and punishing morally deviant behaviours. These ideas were directly linked to the decision to engage the military in Indochina, but were also part of an indirect virility framework related to the more general situation in the region.

This indirect framework essentially used components of the virile nostalgic narrative in the context of domestic political opposition. Not only did the virile nostalgic framework have a direct impact on the French decision, it had other implications not at first envisaged. These various aspects prove that the virile nostalgic framework has more than one application and can be used in various contexts, more or less related to foreign policy. Perhaps the best example of this is found in the debates over the very definition of French virility itself, which helped influence and shape conceptions of what actions to take within the military.

General Conclusion: Which Framework Best Explains the French Response?

This case demonstrates well the importance of employing both quantitative and qualitative methods of analysis. For example, the frequent presence of strategic–material reasoning (52.6 per cent) *does not* prove that realist-strategic ideas were key factors in influencing the decision to go to war. After delving further in the documents, it appeared that (1) these arguments/forms of reasoning were incidental, and evidence of their use in the decision to go to war was present in only 2.8 per cent of the documents overall, and (2) that these arguments were often related to logistical discussions about how to prepare for and conduct the war. Furthermore, securing access to resources went no further than being mentioned in some parliamentary debates and the strategic location of Indochina was not used as a decision-making argument. Moreover, the Cold War argument was found mostly when it was brought up in questions about the influence of the Chinese Civil War on China's international posturing, and not as a reason for engagement in Indochina. On the whole, then, the realist-strategic theoretical framework can be ruled out as providing the best explanation for the French decision to go to war.

The DTW framework did not fare much better (evidenced in only 0.3 per cent of documents overall) and can also be ruled out as a potential explanation. No evidence suggested that France used the conflict to divert domestic opinion from domestic issues. The only notable information found relating to this framework was that the French were united in agreeing that colonial issues were purely a matter for France and not for any of the other (regional or international) actors, including the UN.

Compared to these "old" theories, the "new" ones proposed in this study (status and nostalgic virility) performed well and appeared relatively influential. Status-related arguments were identified in 37 per cent of cases and account for 33.3 per cent of the total number of arguments found in French documents. Nostalgic virility was present in 55.4 per cent of documents and corresponds to a total of 47.7 per cent of the arguments identified. So, we may confidently assert that it was a combination of status-seeking and nostalgic virility – and not just one or the other – that provides the best explanation as to why France decided to go to war in Indochina. Key aspects of the combination were clearly and frequently identified throughout each phase of the conflict.

As status was understood by the French as relational, the need to send appropriate status signals to secure great power status was vital. And, as France linked a civilizational role to great power status, Indochina was a symbolically decisive location. Not only did the French have to send signals of resolve and strength but they also needed to send signals of composure to avoid antagonizing key allies. Virile considerations reinforced the necessity to send appropriate status signals; appearing strong, as well as the idea of avoiding symbolic castration, were of great concern. Primarily, the appropriate signals viewed as necessary to appear strong consisted in punishing morally deviant behaviours.

5

Britain Confronted in the Falklands Crisis, 1982 (I)

Testing Mainstream Theories

2 April 1982 marked the beginning of the Argentinian invasion of the Falkland Islands, an archipelago of roughly 12,000 km^2 claimed by both Britain and Argentina and inhabited by 1,800 British citizens. This provoked an armed conflict with the UK that ended on 15 June of the same year. The conflict ultimately resulted in a British victory and a regime change for Argentina as the military junta collapsed and democratic elections were organized.

Having started as early as 1837, the dispute over the territory was not new. Despite signing a 1971 Cooperation Agreement, an Argentinian destroyer opened fire on a British ship in 1974, leading to a breakdown of diplomatic relations from 1976 to 1980. Britain and Argentina had been negotiating the future of the Islands for the past seventeen years, although despite these efforts very little progress had been made because of the different views held about the administrative status of the Islands and the position of the islanders. The junta grew increasingly tired of the situation, especially as it faced strong domestic criticism of its human rights track record and its poor management of the economy.

The situation spiralled out of control on 18 March 1982 when an Argentinian entrepreneur, Constantino Davidoff, arrived in South Georgia (a British territory administratively attached to the Falklands) with the initial purpose of removing abandoned whaling stations.[1] The conflict began with a problem of documentation: Davidoff requested authorization at the British Embassy in Buenos Aires but did not wait for the response, thus landing in South Georgia without the proper paperwork.

When he refused to leave, Davidoff challenged British authority and jurisdiction over South Georgia and the Falklands. London sent a Royal Navy ice patrol, HMS *Endurance*, to investigate, with instructions to evacuate the Argentines if need be. Buenos Aires responded by sending warships to protect Davidoff's expedition.[2] Faced with growing opposition at home due to his domestic policies, General Leopoldo Galtieri, leader of the junta, took a gamble by invading the Islands in order to boost the junta's popularity.[3] The junta believed the UK would not fight for islands so distant (c. 8,000 km) from London[4] and was also counting on US support.[5] Both assumptions proved wrong.

Three days after the invasion, Britain dispatched a Task Force to reconquer the Islands, including seventy warships and 6,000 elite troops. Initially, the Task Force was sent only to back British negotiations and the UN Security Council Resolution 502, which demanded that the Argentines withdrawal. As Argentina refused to leave, London invoked Article 51 of the UN Charter – the right to self-defence – and the situation escalated into war.[6]

Despite a British victory, the situation has still not been fully resolved to this day. Argentina still claims the Falklands some forty years after the war, to the extent that Britain organized a referendum in 2012 to ask the islanders if they wanted to remain British. The result was 99.8 per cent in favour of British citizenship.[7] As soon as the results were revealed, however, Buenos Aires declared them illegal. These political efforts to bring an end to the conflict follow decades of British military efforts to deter any new Argentine invasion, such as spending over three million pounds in fortifications between 1983 and 1988[8] and permanently deploying fighter squadrons to ensure air superiority in the area (RAF Flight No. 1435).

The Debate over the Falklands Crisis: How Has It Been Studied to Date?

Numerous books and articles, written from an array of perspectives, have been dedicated to the Falklands crisis.[9] Some scholars have suggested that mutual misperceptions between Britain and Argentina led to an escalating crisis that eventually, and inadvertently resulted in armed conflict.[10] Other researchers have suggested that the Thatcher government believed that intervention was necessary to protect democracy and international law.[11] Neither of these explanations dominated the existing literature, however. Instead, the existing literature tends to focus on two dominant explanations.

The first (and perhaps the most enduring and widely influential explanation of the conflict) explains it by reference to the diversionary theory of war (DTW). Scholars in this camp, then have argued not only that, faced with growing opposition at home due to his domestic policies, General Leopoldo Galtieri (leader of the ruling junta) invaded the Islands as a bid to boost the junta's popularity,[12] but also that Thatcher's government decided to use force to retake the islands to divert attention from domestic political concerns and bolster its domestic electoral support.[13]

There are two issues with the existing explanations. First, virtually all the existing studies embody important methodological problems, including being based almost exclusively on data (e.g., memoirs, public speeches) that cannot be viewed as highly reliable sources of the internal decision-making processes of key figures. Also, using interpretive methods that do not allow for a systematic and rigorous comparison of the relative weight of the various arguments and strategies of persuasion. Moreover, the conclusions of public, unclassified documents – namely newspapers and House of Commons debates – have to be taken with relative caution: by studying public documents one cannot prove the motives that drive an action; one can only present the public justifications made to intervene. It is for this reason that the analysis of declassified documents takes on such importance, as they allow for insight into the minds and subjectivities of the decision-makers involved.

Secondly, while there are a few studies which examine the role of pride and status in this conflict,[14] none of them offer a rigorous examination of other "emotional" investments and considerations (including many associated with nostalgic virility) that could have played a role in Thatcher's decision to go to war.

Explaining the British Decision: Support for the Theoretical Frameworks

Before turning to the analysis of British decision-makers' perspectives and deliberations over the decision to go to war in the Falklands, I will provide a brief overview of a priori evidence from past literature to explain why the three theoretical frameworks being tested all appear to be relevant explanatory factors.

Status and Nostalgic Virility: A Priori Evidence

Status-Signalling. Since the Suez Canal fiasco in the 1950s, Britain had experienced a succession of embarrassing status issues.[15] Before Suez, it believed itself to be on the same influential level as the US and the USSR.[16] Suez, however, changed everything. The image of the UK as a great power collapsed.[17] The embarrassing combination of military and economic weakness, military and diplomatic humiliations, and a general British disengagement from world affairs had made foreign diplomats ask their British counterparts the question: *What has happened to your country*?[18] In the wake of such events, the arrival and sheer speed of the Falklands invasion has led some commentators to claim that "the humiliation of Britain was sudden, and complete."[19] This idea of humiliation offers a helpful transition to the second component of the model: a nostalgic virility.

Nostalgic Virility. One can find evidence of nostalgia and virility in Thatcher's position even before the invasion of the Falklands. A key component of nostalgic virility is the refusal to appear weak or indecisive, examples of which can be found in various discussions. When talking about the unions with the Deputy Secretary to the Cabinet, Sir Robert Wade-Gery, Thatcher commented: "The last Conservative government was destroyed by the miners' strike. We'll have another one, and we will win."[20] She also appears on many occasions to show strength and resolve by her unwillingness to compromise. Charles Moore highlights several such examples, including her responses to media questioning on whether she would slow down on her economic agenda: "to those waiting with bated breath for the favourite media catchphrase, the 'U' turn, I have only one thing to say, 'You turn if you want to, the Lady's not for turning.'"[21]

Finally, nostalgic virility assumes that leaders compare their country's status and strength to the past, especially to times of glory. Such elements are found when examining Thatcher's behaviour surrounding the 1981–82 economic agenda, when she told an American reporter, "If we had ever looked at Dunkirk as a kind of balance sheet, as sometimes I am asked to look economically at this country, well I don't think we would have gone on at that time. If you looked at it as a matter of spirit of the people, then it is totally different."[22]

Realist-Strategic Frameworks: A Priori Evidence

One a priori argument that supports the realist-strategic interpretation of the British decision to go to war to reconquer the Falkland Islands can be found in the Islands' geopolitical significance. The Islands are located 780 km from Cape Horn and 6,300 km from the Cape of Good Hope. In the eventuality of the closure, for any reason, of either or both the Suez and Panama Canals, the Falklands would be a useful strategic location providing facilities to control commercial lines going through the Capes.[23]

They are also in a good position to control the Strait of Drake, an undersea passage traditionally used by Soviet SSBNs to travel from the Atlantic to the Indian Ocean.[24] In the context of a renewed East-West confrontation, the control of the Falklands would be an advantage. At the time, Thatcher and US President Ronald Reagan were leading the fight against communism and reconquering the Falklands would be a signal to the USSR of the West's resolve just a few years after the Soviet invasion of Afghanistan and of Communist-supported revolutions in Central America – a point mentioned directly by Thatcher in her memoirs.[25] On top of this is the material resources argument, which was uncertain but may well have played a role: The Islands were said to have large oil resources, although the exploitability of the deposits remained to be assessed.[26]

DTW Frameworks: A Priori Evidence

There are three a priori arguments to justify the idea that British motives were driven by domestic political concerns.

(1) *The domestic situation in Britain at the time of the conflict*. The country had been hit by a major economic crisis – the miners' strike – that paralyzed it to such an extent that in January 1982, the government's approval rating was at only 24 per cent. This increased to 48 per cent in June of the same year and the Conservatives ultimately won the general election of 1983. At face value, then, it is quite plausible that the Falklands crisis saved Thatcher's political career.

(2) *Division in the cabinet*. Moore gives various accounts of disagreements between Thatcher and her ministers. For example, she was

opposed to Lord Peter Carrington over how to settle the Falklands dispute – the latter of whom favoured negotiations even after the Argentine invasion – as well as over what behaviour Britain should adopt with respect to the USSR.[27] These divisions were likely important threats to her political stability: "Even before the Falklands crisis, she was on probation with the [Conservative] Party, and had never been properly accepted by many in its higher echelons."[28]

Thatcher and her government's position were put under the microscope and faced accusations of amateurism in the way they managed the crisis. It appeared that the cabinet was well aware that the Argentine Navy, which was dominating the junta, was historically in favour of military action to take the Islands back.[29] The FCO was also made aware by the British Ambassador in Buenos Aires, Anthony Williams, that some newspapers were calling for a renewal of what India did in Goa.[30]

(3) *The strong lobby the Falkland Islanders presented to the British Parliament.* The Falklands lobby was formed in 1968 in reaction to the potential transfer of sovereignty over the Islands to Argentina.[31] Several individuals, companies, islanders, and even some MPs from the three main parties formed part of this lobby.[32] In 1968 it proved its clout by successfully blocking Lord Chalfont's efforts to transfer the Falklands to Argentina and to convince the islanders that this was the best course of action for them.[33] By 1979 the lobby was still a powerful force, as indicated in a private note from John Ure[34] to Nicholas Ridley[35] stating that it had enough power to get politicians elected or remove them from Parliament.[36] Accordingly, it is plausible that the lobby's political influence was another consideration Thatcher faced in weighing her options and confronting the future prospects of her own political career.

As the above overview of a priori evidence displays, all three theoretical frameworks considered in this study pose strong possibilities for explaining the British decision to go to war in the Falklands. Evidence for both the realist-strategic and the DTW frameworks is presented in this chapter, followed by a presentation of the evidence for the status and nostalgic virility frameworks in the next chapter.

Evaluating Realist-Strategic Arguments: Little Evidence, Many Politicized Narratives

Overview of the Results

Analysis of the dataset revealed that some 50.1 per cent of documents contained discussions related to strategic and material considerations. A first impression of this frequency may lead one to believe that the realist-strategic framework is thus dominant. However, a close qualitative reading of these documents shows that the substantive importance of these discussions in the decision to go to war over the Falklands is actually much lower than it appears. The quantity of 50.1 per cent is misleading on its own as it captures not only the key hypotheses of the realist-strategic framework (i.e., regarding decisions to go to war) but also broader descriptions of the situation that use strategic-material terms (i.e., descriptions which give us insight into the broader understanding of the conflict and its context by decision-makers). When leaving these descriptions aside, we find a completely different story: *the core of the realist-strategic framework is present in only 3.6 per cent of documents*. Despite being so low, it is still worth analyzing these scant findings through a qualitative lens to reveal some interesting and nuanced aspects surrounding the conflict.

Economic Interests ... That Turn out to Be in Favour of Abandoning the Islands

Gustafson[37] and Gibran[38] both suggest that the exploitation of oil near the Falklands cannot have been considered as a serious argument. My findings support this conclusion. Among coding questions designed to capture real economic concerns, only two were present in the dataset: *concerns about potential trade gains* and *concerns about profits for national companies* (both found in only 0.7 per cent of documents). These arguments appear only in the first phase of the conflict[39] and disappear completely after 21 May 1982 (the date the British commando brigades landed in the Falklands).

Where identified, economic arguments were in fact used to *support* the option of negotiating over sovereignty with Argentina. On 21 September 1979, Thatcher was told that "a settlement [over sovereignty] with the Argentines would open the way to development of a 200-mile fishing zone ... It would also greatly improve prospects for Anglo-Argentine trade … [It] might strengthen our eventual prospects of preferential access to potential large minerals and hydrocarbon deposits there."[40] In 1980, Nick Ridley, wrote to the Foreign Secretary, Carrington, that if the dispute were settled "our relations with Argentina would improve, leading to increased commercial and arms business and the opportunity to start exploiting the resources of the South West Atlantic area."[41] In September 1981, Thatcher was told that it would be best to find an agreement with Argentina because it was the country providing the most services to the Falklands (e.g., flight connections, basic needs, etc.), and if this were to stop, Britain would not be able to cover these areas.[42] In sum, if there were any economic advantage for Britain, it would most likely have been to withdraw from the Islands.

One final element of the economic argument was that in the eventuality the Islands would have to be defended, such an operation would be too costly. In 1981, many contingency plans were evaluated as negotiations with Argentina came to a dead end. These plans considered and compared having to supply the Islands directly from the UK versus permanently dispatching an aircraft carrier strike group for deterrence. Ultimately all plans were too expensive. Even in the case of an invasion, any operation for reconquest was seen as too harmful to vital UK interest.[43]

Instrumentalizing Strategic Interests: Two Different Stories

Analysis reveals that the situation was often described in strategic terms (29.4 per cent of documents), but when we look at the realist-strategic framework of deliberation in detail the findings become much more nuanced. How was the realist-strategic framework organized? What idea(s) is it built around? What does it tell us about the decision to go to war?

First, let me be clear about what it does *not* tell us. I found no evidence to support the Cold War component of the realist-strategic framework. To test for the Cold War hypothesis four argumentative logics were investigated: (1) *the domino effect,* (2) *rolling back the adversary,* (3) *potential loss/gain of buffer zones,* and (4) *potential loss of bridgeheads.* Of these, only the *domino effect* was identified, and in only 0.5 per cent of documents. Rather, the analysis revealed that *concerns about gain/loss of strategic locations* was the only meaningful argumentative logic found to support the realist-strategic framework, appearing sparingly in 1.9 per cent of documents. A qualitative reading of these documents reveals two distinct stories: the symbolic importance of the Falklands as a strategic location and the narrative portrayal of the US as a "villain" for its lack of support.

Fighting over the Falklands to Secure Gibraltar and Hong Kong

Evidence suggests that members of the cabinet were worried about the implications of the Falklands issue on the British possessions of Gibraltar and Hong Kong. Concerns over the gain/loss of strategic locations appear in 2.2 per cent of ministerial documents, particularly those meant to remain within the government and its administration. Some ministers thought that negotiating over the sovereignty of the Falklands would create issues with Spain concerning Gibraltar,[44] and might suggest to China that Britain was ready to surrender Hong Kong.[45] These concerns also appear in some 2.6 per cent of Thatcher's documents. At face value, these findings appear to lightly support the prospect theory argument.[46]

Evidence supporting the prospect theory is found from ministers in *Phases 1* and *3* (2.9 per cent and 2.3 per cent, respectively) but not in *Phase* 2. In terms of support for the theory, this distribution is quite problematic. *Phase* 2 corresponds to the moment when the decision to send the Task Force was taken, so it is at this moment where most of the motives should have been explicitly expressed. The analysis shows no evidence of this. Furthermore, in the documents analysed concrete concerns other Gibraltar never went further than Thatcher "ask[ing] whether all necessary precautions had been taken against a possible military attempt on Gibraltar."[47]

The US as "Villain" in a Hero-Villain-Victim Narrative Constructed around Strategic Arguments

Britain's dominant material-strategic concern, as outlined above, was the gain/loss of strategic locations. What is interesting from a qualitative perspective (and despite the low frequencies) is that these concerns appear primarily in external documents, and particularly in those issued by Thatcher, such as in conversations and letters with US President Ronald Reagan.

At the beginning of *Phase 3*, US support for the British retaliation was far from certain. Reagan, through US Secretary of State Alexander Haig's proposals, favoured a diplomatic solution and tried to stop

Thatcher going to war. In fact, the US thought that supporting the UK would undermine its position in South America as Argentina was a strong ally in their fight against communism.[48] To be successful, the British operation needed US assistance in intelligence gathering, support for its fleet, and permission for the Task Force to use the American facilities on Ascension Island.[49] Even with the island in British hands, the military facilities belonged to America and their use required US approval.[50] Evidence in the dataset shows that gaining US approval for use of the Ascension Island facilities was rather difficult to obtain. To avoid infuriating South American countries, the US declared that the use of their facilities by the British would be restricted to a bare minimum. This infuriated Thatcher, although she ultimately obtained a satisfactory outcome: the island would be used but any mention would disappear from Haig's official speeches.[51]

To gain the much-needed US support, Thatcher used a combination of two material-strategic argumentative logics. First, that it was in the US interest for Britain to win the conflict. Second, that by refusing to clearly support Britain the US was actively fighting against its own interests. To make these arguments, Thatcher drew Reagan's attention to the strategic importance of the Falklands. During her phone call to him on 13 May, as fighting was intensifying in the wake of both British and Argentinian ships being sunk (the most famous being the Argentine ARA *General Belgrano* and the UK destroyer HMS *Sheffield*), and just one day after additional landing forces departed from Britain, Thatcher presented British possession of the Falklands as being vital to US interests: "The Falklands are important strategically. You've only got to have the Panama Canal closed and that would be easy. You need a passage through the Magellan Strait … you need Port Stanley, and you need it in hands of a usually friendly power … And it was important during the last war to the Western world, and it was very important in World War One too."[52] During a dinner conversation with Haig on 8 April 1982, she posited the idea that by not giving their support the US was in fact reinforcing the USSR's position – and thus seriously undermining vital US interests – as the "military junta [is] in bed already with the Soviet Union."[53] Despite this context, the frequency of arguments connecting the Falklands to the fight against communism was very low (1.5 per cent of documents).[54]

Linking the Falklands crisis to US strategic interests was only one aspect of Thatcher's discourse. Another was the inclusion of a hero-villain-victim narrative in which the victim was Britain, the hero was the British military, and the villain the US. The US had disappointed the British government by remaining extremely cautious and by making peace proposals through Haig.[55] The proposals were mostly centred around the idea of stopping and then withdrawing the fleet, calls for a UN administration of the Islands, and possible talks over sovereignty.[56] Thatcher told Haig about her dissatisfaction as early as 8 April 1982,[57] when the first warships were gathering in Ascension Island. However, the best examples are found in her discussion with him on 14 April (the day the first group of ships departed from Ascension Island to the Falklands): "It's our island and the dispute is about our island," she stated. Refusing to entertain the idea that Britain should be forced to agree on an unsatisfactory settlement, she further complained: "We are not getting any help from our traditional allies."[58]

Thatcher expressed her dissatisfaction directly to Reagan, as well. "I cannot conceal from you how deeply let down I and my colleagues would feel if under those circumstances the US were not now to give us its full support."[59] Evidence of her distrust can be seen during the Defence and Overseas subcommittees of 19 and 22 April, when she ordered that the US should not be informed of any British military plan or operation in case they were to leak them to Argentina.[60]

The dissatisfaction expressed was proportional to the need for US support and the authorization to use Ascension Island facilities. However, the narrative changed once the US finally decided to fully support the UK, as marked by the appearance of narrative rhetoric portraying the US as heroes on 4 May, when she concluded her letter to Reagan by writing, "Finally, let me thank you once again for your splendid support, it will make all the difference."[61] It was a complete shift in tone.

Conclusions on the Realist-Strategic Framework

Despite the active search for both explicit and implicit arguments linked to the various material-strategic and realist explanatory frameworks, no tangible results of significance were found. To be sure, the frameworks are present in the description and framing of the situation and of available options, but no evidence is found to suggest that these considerations can explain the British decision to go to war and reconquer the Falkland Islands. Accordingly, this lack of evidence and explanatory power directs us to consider other potential frameworks for explaining the British decision to go to war.

Evaluating DTW Arguments: A Tiger Paper?[62]

If realist-strategic arguments and the British overseas interests fail to account for much of the argumentative logic or decision-making leading to the choice to go to war over the Falklands, then what about domestic British interests and concerns? The remainder of this chapter will evaluate the evidence for DTW approaches and explanations of the British decision to determine whether the domestic political context and Thatcher's supposedly precarious political career indeed constituted primary decision-making factors.

Ambiguous Results: A Framework Present but Not Where It Was Expected

Analysis revealed that elements related directly to the DTW framework were present in some 13.6 per cent of British declassified documents. However, when also including broader occurrences of terms, rhetoric, and expressions focused especially on domestic, legal, or political issues, the frequency rises to some 18.2 per cent of documents. These numbers point to the presence of a clear domestic aspect in the decision-making surrounding the conflict, meaning that, at the very least, it formed part of the bigger picture of British motivations and provided narratives worth investigating further through qualitative analysis. Before moving to the qualitative analysis, however, a deeper look at the frequency distributions will help to contextualize and identify potential narratives.

Phase 1 began with relatively strong evidence of the DTW framework (26.9 per cent of documents) but in fact decreased to 11.6 per cent at *Phase 2* and to only 4 per cent at *Phase 3*.

The DTW is very unlikely to explain, by itself, why the British government was adamant to continue its move towards the Falklands despite calls from other countries, including the US, against such a move. The framework predicts that leaders go to war to save their political careers: it was therefore necessary to find constant and strong concerns made by individuals about their political future. However, the results show

Table 5.1 Phase distribution of evidence for DTW

Phase	*Per cent of cases*
1	26.9
2	11.4
3	4

Table 5.2 Distribution of evidence supporting DTW by actors

Actors	*Per cent of cases*
Thatcher	6.8
Ministers	24.1
Public (MPs, domestic opinion, Islanders)	5.1

that the situation was less clear-cut than the ideal depicted by the DTW framework: while the ministers' score (24.1 per cent) reflects the framework's assumptions, Thatcher's score (6.8 per cent of documents) goes against them.

Thatcher's score is surprising in light of various arguments in the literature explaining her position as precarious, and that the invasion "posed a deadly threat to everything she valued, not least her own political future."[63] Based on these assessments, there was an expectation of finding more evidence from her to support DTW. Thatcher's score seems to disprove the idea that the British decision was motivated by her will alone in order to save her political career. The most compelling explanation for this lack of evidence is that ministers themselves possibly pressured Thatcher to act, threatening to challenger her if she refused. In this case, the explanatory power of DTW is orientated toward the cabinet, consistent with the claim that "the most pressing concern driving the government's policy to begin with was its own political survival."[64] I did not find such evidence in the data.

At the very least, then, we could argue that DTW was one motivating factor among others for ministers, and that there was at the very least a narrative developed which centred around the idea of remaining in power. While the bulk of evidence for DTW comes from ministers and decreases through each phase of the conflict, we find a more nuanced story of Thatcher's concerns for her political future when analyzing the findings based on type of document. While 37.5 per cent of Thatcher's internal documents from *Phase 1* showed evidence for DTW, no such evidence in internal documents was found for *Phases 2* or *3*. However, in *Phase 3* some 13.3 per cent of external and 4.2 per cent of public documents from Thatcher displayed evidence for DTW. These results raise the likelihood that concerns over losing power were not genuine for Thatcher, but rather used externally and late in the conflict to gain support from other countries (i.e., after the decision to go to war was already taken).

Overall, the quantitative results show that the DTW framework is present and much more consistent than that of material-strategic arguments. But how representative is it? To what degree does the DTW explain the British reaction? To what extent is it important in the decision-making process? To answer these questions, the data was coded to look for any references to domestic concerns, ranging from concerns about an election to the need to divert public opinion, as well as the presence of any implicit or indirect mention of these ideas. A qualitative reading of these findings paints a clearer picture of the significance of DTW in Britain's decision to go to war.

A Diversionary War to Hide Criticisms of Governmental Carelessness?

Some observers have suggested that the British government went to war to conceal its mistakes and supposed carelessness in issue management.[65] My initial finding was to the contrary: the British government was not in fact that careless about the Falklands issue. Two codes were developed to assess this claim: (1) *seriousness of the issue/issue difficult to solve*, which tracked each explicit reference to whether the issue was, would, or should be considered a government priority; and (2) its negative counterpart, *issue not very important/less important than another one*, which tracked each explicit reference that it was not, would not, or should not be the case.

The first code was found in 13.1 per cent of documents and the second in only 0.7 per cent. Taking a closer look, Thatcher was the only one to explicitly demonstrate any significant evidence of carelessness.

However, this isn't very significant as it came from just one document, dated from 15 October 1979 – the year she took office. During a Defence and Overseas Committee meeting, she said that the "discussion of the Falklands should be postponed until after the Rhodesian issue has been settled."[66] Ministers, on the contrary, attributed considerably more significance to the Falklands issue (13.9 per cent of documents treating it as important compared to only 0.7 per cent as unimportant).

Furthermore, discussions in cabinet demonstrate a careful – not careless – consideration of responses to the Falklands issue. As it became clear that Davidoff would not leave South Georgia of his own accord and that he was getting support from the Argentinian junta, the possibility of sending a task force to the Falklands was discussed in cabinet. Several options were outlined but all were rejected for fear of adding fuel to the fire. The first option, rejected by both the FCO and the Chief of Defence Staff, was to send a group of seven destroyers. The former argued that it would be counterproductive[67] while the latter suggested the option was not viable because this force would have no air cover.[68] The option was, therefore, abandoned.

Another option was to send a much stronger task force, "comprising a carrier, at least 4 or 3 surface ships, possibly a commando group of 800 Royal Marines, and perhaps a submarine."[69] This option was considered worse than the previous one: such a force could easily be spotted, which would in turn surely increase tensions and accelerate Argentine moves (if any had been prepared) to invade the Falklands. Both the FCO and the military warned that doing so would lead to an invasion.[70] The deployment was considered as effective only in the eventuality of a possible invasion, but as "there is at present no sign of this," the option was not carried forward.[71] From this point of view, it would have been careless to send part of the fleet in the knowledge of the potentially catastrophic effects of such a decision. Instead, it was carefully decided that two nuclear submarines would be sent to the South Atlantic: they would be more discreet, thus having the double advantage of protecting the Islands and strengthening negotiations.[72]

Overall, both the quantitative and qualitative analyses demonstrate that the Falklands issue was carefully considered and discussed by Thatcher and her cabinet. These findings oppose charges of carelessness and the concealment of mistakes noted in some of the literature.

British Use of the Losing Power Argument: Various Narratives

Concerned for Their Political Fate, You Say? Possible, but Not Very Likely

The idea developed in the literature that the British government was careless is not the only claim challenged by this analysis. The results obtained also challenge the idea that there were strong concerns about elections and that Thatcher and other ministers were concerned about their political future. Explicit and implicit concerns about elections were found only to be about local ones and were exclusive to MPs (2.6 per cent of cases in MP documents), due mainly to the economic repercussions the conflict would have on their local constituencies.[73]

In fact, despite an active search for references to the need to divert domestic opinion or to rally it around the flag, none were found. What I did discover, however, were some concerns expressed by Thatcher about parliamentary hostility (6.8 per cent of her documents) and party hostility (0.9 per cent). They were mainly apparent before the Argentine invasion. Conscious that percentages can be misleading,[74] I retrieved all DTW-related evidence coded in *Phase* 2, where, in theory, any decisive evidence in support of this framework should most likely be found. There were fourteen instances in total, with only four related to cabinet discussions on 2 April (the day the decision was taken to assemble the task force) and 3 April (when the last outnumbered and outgunned marines at Grytviken surrendered). The content of these instances is, however, far from decisive.

During cabinet meetings on 2 April, it was said to be expected and even "normal" to face criticism, and that there was nothing to worry about. Criticism, particularly in Parliament, that the government had been taken by surprise and had failed to act early enough was to be expected. It was pointed out that the origins of the crisis lay in the landing of the Argentine scrap metal merchants in South Georgia only twelve days earlier; there had been no reason at that time to think that this had been engineered as a prelude to the invasion.[75]

However, this image of a seemingly "relaxed" government in the face of parliamentary critics could well have been fabricated. Two other instances found were in Ian Gow's (private secretary to the prime minister) notes on the 3 April cabinet discussion, which suggested that the government was much less in control of the situation than the minutes indicate. His notes reveal that the meeting was rather heated and that some members described the situation as "hideously embarrassing."[76] Would this revelation cast evidence in favour of DTW? Probably not. The comments were made by conservative politicians who had little influence and were hostile to the cabinet. What it does suggest is simply that the content of the minutes may differ from what was actually said during the meeting, as the official version often emphasizes only what was agreed. Whatever the case, if there were fears of losing power at this time they were not captured or formally expressed.

A potential divide related to DTW was identified between Thatcher and her cabinet about what course of action to follow regarding negotiations with Argentina. She favoured the option of not transferring sovereignty to Argentina in 24.8 per cent of her documents. She used the argument of parliamentary and party hostility to oppose Carrington's suggestion of discussing a leaseback option over the Falklands. Her first opposition to the option can be found as early as 1979 when she annotated a report from Carrington: "I could not possibly agree … Nor would it even go through the HMC and the Conservative Party."[77] One year later, at a Defence and Overseas Committee meeting, she argued that the leaseback option should be abandoned: "Parliament's hostility to the idea of a settlement was now so strong that even if the islanders' considered view was that one [leaseback to Argentina] … this might be regarded as merely the result of pressure from the Government."[78] In fact, among all the outcomes possible before the invasion – independence, Fortress Falklands, negotiation with Argentina, freezing the dispute/keeping the islands under British rule, and leaseback – she remained strongly in favour of freezing the dispute (found in 47.5 per cent of Thatcher's documents in *Phase 1*), although she did allow the FCO to pursue negotiation with Argentina. Among her favourite expressions at this time was that "the wishes of Islanders are paramount."

By contrast, the cabinet appeared much more divided over which policy to follow. In ministerial documents drafted before the invasion, calls for negotiating with Argentina appeared in 33.7 per cent of cases, freezing the dispute in 11.6 per cent, and leaseback in 14.5 per cent. Those especially in favour of negotiation and leaseback included Ridley and Carrington, as noted in the literature. It is in this context ("which option should be followed?") that concerns about lobbies first appeared (5.8 per cent of documents). Parliament was against the idea of leaseback and demonstrated its hostility to Ridley and Carrington's ideas.

Parliamentary hostility grew stronger at the news of the 1981 Defence White Paper's suggestion to withdraw HMS *Endurance* from the Falklands in order to save costs.[79] As a result, evidence of concerns over parliamentary hostility appeared in 10.2 per cent of documents.[80] Not only was there parliamentary hostility to manage, but also the framing of the government as the "villain" of the story (found in 9.4 per cent of MP's documents – the exact same frequency of "villainous" charges levelled against Argentina). Overall, then, we find that electoral politics were almost inexistent as an influencing factor in Thatcher's deliberations, and that very few other actors close to Thatcher expressed consistent concerns over their political future. Rather, concerns turned largely around party and parliamentary hostility, which were often swept aside as expected consequences of any slow, well thought-out government response to the crisis.

The US as "Super Villains": "Pretend" Domestic Difficulties and the Attempt to Secure Vital US Support

During the analysis of the data, I found a narrative using the idea of being on the verge of losing power. This narrative was identified in Thatcher's external documents, in which the idea of parliamentary hostility appeared in 8.3 per cent of cases. This evidence was found during her talks with Haig, as the latter was shuttling between London, Washington, and Buenos Aires trying to find a compromise that would avoid military confrontation. For Britain, this was a crucial moment because it had an urgent need for full US support. As analysis reveals, the DTW framework was used by Thatcher in conjunction with the realist-strategic framework to gain US support.

The first time she used the DTW framework was at a dinner with Haig on 8 April (three days after the fleet's departure), when she argued in support of the notion of a "villain": "Ministers would face a motion of censure [if] British administration was not restored following the pledges which the Government had given in the House of Commons." She concluded: "If the Government had not been able to announce in that debate that the fleet was to sail, the position could probably not have been held."[81] By not helping her government, the argument went, the US was creating a situation for a change of British government that might have serious implications for US interests – notably in their fight against communism. The argument of losing power because of a lack of

US support came again in Thatcher's discussion with Haig a few days later, when the fleet was nearing the Falklands and the US was pushing her to accept a compromise.[82] She told him on 12 April that if the fleet were to stop in the vain hope that Argentina would fulfil its engagement to withdraw from the Islands completely, "she would not survive."[83]

Two days later, on 14 April (the day the fleet departed from Ascension Island for the Falklands), the narrative that the US was the villain of the story and that the government risked falling if the US did not support Britain continued to develop. By this time, UK public opinion toward the US was being included in the narrative: "Part of your statement will have an extremely adverse reaction here … [because] there's a feeling that you're trying to put the screws on us to make it difficult for us to liberate our Islands from the aggression of a military junta who's in bed already with the Soviet Union."[84] The link with the realist-strategic framework was made: the US, barring support for the UK, are the bad guys – worse even than the Argentines. Not only did US behaviour go against their vital interests, not only would it be playing the USSR's game, but it would be siding with a dictatorship to such an extent that a political situation would be created that could result in the fall of a British government that was possibly the friendliest of all toward the Reagan administration. To be fully complete, however, the hero-villain-victim narrative requires a set of goals to be pursued by the hero. With little to no evidence for such goals found through investigation of realist-strategic and DTW frameworks, the nature of these goals will be investigated in the next chapter by testing the theoretical frameworks status and nostalgic virility against the data.

Conclusions on the DTW Framework

Thus far, this analysis has shown that there is some evidence of the DTW framework. However, the question of whether this framework is best suited to explain the British decision to invade the Falklands remains to be seen. It would appear, certainly in Thatcher's case, that claims in the literature are mistaken: there is no real evidence that Thatcher was concerned about her election prospects; by contrast, ministers seem to have been the ones more anxious about their political careers. However, these concerns remain marginal overall, based on both quantitative and qualitative readings of the data. If the DTW was not a key motive for intervention, it played an important role nonetheless – especially by Thatcher – as a form of diplomatic leverage in discussions with the Reagan administration.

Based on these findings, should we rule out the possibility of a more important role for DTW in the British decision? This would seem presumptuous. Politicians may have thought about saving their political careers without expressing these views outright, although without access to an individual's mind it is impossible to say for sure. Rather, it would make more sense to assume that concerns about parliamentary hostility caused the government to act slowly and present their plans for the Falklands unclearly, thus causing divisions within government. At the very least, the claim can be made that the DTW framework was present and was part of the arsenal of arguments given by the British to secure vital US support. However, the data does not support an interpretation of DTW as constituting a primary cause for the British decision to go to war in the Falklands.

6

Britain Confronted in the Falklands Crisis, 1982 (II)

Testing Status and Nostalgic Virility

This chapter continues the investigation of British motives for going to war in the Falklands by testing the frameworks of status and nostalgic virility against the data. Here, the empirical evidence for these frameworks will be evaluated and placed in conversation with the evidence for realist-strategic and DTW frameworks outlined in chapter 5. We will begin with an evaluation of evidence for the status framework, looking at questions of great power status and status signalling, followed by an evaluation of evidence for the nostalgic virility framework. By the end of this chapter, we will have a more holistic understanding of the perspectives and motivations leading British decision-makers to engage in war over the Falkland Islands.

Considerations on Great Power Status: The Symbolic Aspects of the Conflict

To continue our investigation into the evidence for status and nostalgic virility as explanatory factors, we will outline the symbolic aspects of the conflict and how status perceptions play a key symbolic role in international conflicts more generally. To begin, let's first consider the overall quantitative results to understand how frequently status considerations appeared in British decision-making documents.

Overview of the Results

Overall, evidence specifically supporting arguments within the status framework is found in some 17.6 per cent of documents, although this frequency increases to 42.9 per cent when including broader references

Table 6.1 Distribution of status-related ideas

Phase of conflict	*Status per cent*
1	24.8
2	11.6
3	14.7

Table 6.2 Distribution of status codes among actors

Actor	*Per cent of cases in dataset*
Thatcher	23.1
Ministers	16.1
Public	28.2

to and considerations of status that don't fall directly into the status framework yet nonetheless provide insight into the contexts and perspectives of decision-makers. These numbers suggest a firm place for status considerations within the ecosystem of influences on British decision-makers. So, who was most concerned with status? In which type of documents do we find the majority of status arguments, and at which phase of the conflict?

Status concerns should in theory be constant, as the framework emphasizes the need to send appropriate status signals to keep great power status.[1] We should therefore find relatively high levels of evidence in *Phases 1* and *3*, while in *Phase 2* (when the decision to send the task force was made) any status-related arguments should be present in the form of debating concrete options for signalling the status claim. The specific ways in which these arguments may have been framed is to be qualitatively investigated, but for now let's consider the overall presence of the framework and its distribution. As table 6.1 shows, status considerations are present at a fairly significant yet declining frequency throughout the conflict, but with a steep decline in *Phase 2*.

Regarding the expression of status-related ideas by individuals, it appears that all actors investigated viewed and framed the situation in explicitly status-related terms (12.4 per cent of cases overall), but that those most concerned with status were Thatcher and those within the category public (e.g., MPs, the press, domestic opinion), as summarized in table 6.2.

This distribution reveals a very interesting story. There seemed to be, at most, an opposition and, at least, a clear divide between Thatcher and her cabinet ministers. Whereas the evidence for DTW showed that ministers appeared to be more concerned than Thatcher by the prospect of elections, here we find the reverse: Thatcher is the one who is more concerned by status issues.

The second result that the distribution suggests is a revival of an argument connected to DTW that recalls the Roman practice of evergetism (which was not actually tested for DTW because it was off the radar of theoretical assumptions).[2] As public opinion was strongly concerned with the country's status and ranking on the international stage, we could make the argument that, knowing this, Thatcher and her cabinet decided to go to war to give the public a display of strength and prestige (if victorious). The quantitative results signal in favour of such an interpretation. Status-related ideas/arguments were far more present in *external* (29 per cent) and public documents (29.9 per cent) than in internal ones (11.7 per cent). These numbers could indicate that Britain publicly intended to fulfil its status-related role by intervening in the Falklands.

Indeed, this seems probable in light of the fact that many external documents signalling status concerns/perceptions were issued by the British government and directed to other heads of state and government, while the high level of status-related public documents suggests that Thatcher and her ministers were sending signals appropriate to domestic opinion. How can we interpret these numbers? Theoretically speaking, by publicly reinforcing status signals they would show that there were no worries about British status, that they were well aware of Britain's high ranking in the world, and that they were committed to keeping this position by going to war over the Falklands. We can expect them to have given the public signs that they take Britain's status legacy seriously. In this case, we would need to find within the public documents evidence of status-signalling and evidence of "selling the war" to the public. These particular themes are investigated below. First, however, let's review what great power status really meant for Britain at this time.

A Consensus among Actors: Britain's Particular Status

Great Power Status: How Is It Conceived by the British?

The academic literature provides various "great power" definitions, but the question remains as to whether such ideas are a purely academic trope or whether they are shared by decision-makers. To try and answer this question, I coded the data for every explicit reference to ideas defining what it means to be a great power. Such references were quite sparse (2.4 per cent of documents) but interesting in qualitative terms. Indeed, several key characteristics of great power status were found.

Through detailed examination, three main ideas were identified that made up the bulk of interpretations of Britain's great power status:

(1) *A strong military.*[3] This is the basis for any great power as it enables a country to pursue its foreign policy objectives and safeguard its interests. The very low frequency of this idea is quite surprising; it was mentioned only twice. One was when Cecil Parkinson (Junior Trade Minister and member of the small war cabinet) wrote a letter to the Conservative Party giving reasons why Britain needed to react to the Argentine invasion.[4] The other was in a cabinet minute in which Thatcher praised the "professionalism and skill" of the Armed Services.[5] Basically, the argument for military strength was based on the need to show strength because Britain's power had been forgotten, and the need for the military to be seen as efficient and admirable so that its strength is recognized.

(2) *Having interests outside their own region and proving their global influence.* In the dataset, such mentions were not explicit and rarely mentioned by British decision-makers. I found references, although not many, to Gibraltar and Hong Kong, the latter being used as a possible inspiration point when looking for a solution to the Falklands sovereignty issue.[6] Other mentions were concentrated on the British position in the South Atlantic, including a British survey in the Antarctic.[7]

(3) *Managing the world order* (e.g., enforcing international law, solving crisis, etc.). This argument was found in 2.2 per cent of the dataset. The basic notion is that conflicts and crises should be managed by the most powerful and influential countries in the world. To be effective, laws must be backed by force. This dynamic was clear in

> the Falklands conflict. Argentina organized an invasion to claim a territory because it felt that negotiations were ineffective to regain what it believed was rightfully theirs. If this sort of behaviour remained unpunished, it would set a precedent implying that the respect for international law was counterproductive. Thatcher's argument to Reagan was that Britain could not let the situation remain unanswered because it would potentially be too dangerous a precedent.[8] These arguments are backed by the presence of other related codes such as those measuring the role Britain believed it had on the international stage, namely as protector of democracy and international law. I will return to these roles later.

So, do these characteristics of great power status have any direct link with the British decision to react with military force to the Argentine invasion? Although the frequency of their use is not high, the answer appears to be yes. The argument for great power duties occurs in 5.8 per cent of documents as an explanation for reacting with military force, while another 2.4 per cent of documents contain direct references to the various components of great power status outlined above. The subtle yet important differences between these two argumentative logics (i.e., the duties and nature of great power status) were found to drive distinct narratives, both of which were reinforced by frequent indirect references to such ideas. The details of these narratives are outlined in the next two sub-sections.

A British Feeling of Superiority Based on Political Regime and Military Strength

Explicit mentions of what rank Britain had in the international hierarchy – namely, great power status – were rather low.[9] However, the broader idea of Britain having a specific status or role in the international hierarchy was found in some 29.3 per cent of the dataset, with a peak at 40.2 per cent in documents written by Thatcher. These ideas were present, albeit at a lower rate, in ministerial documents (24.8 per cent) and domestic opinion documents (35.9 per cent). In addition, 6.3 per cent of documents demonstrated evidence for a feeling of superiority, the bulk of which came from domestic opinion. The ideas of having great power status and of being a superior country were distributed across the various phases of the conflict at a relatively high frequency.[10] What differs are the very precise ideas linked to these argumentative logics at each phase.

In *Phase 1*, both ideas were present at relatively similar frequencies. Feeling superior, at this phase, was built on the feeling of having superior political institutions and being on the side of the law. Both aspects were present in all types of documents, although the most explicit examples were in *external documents*, especially during British-Argentine talks. For example, in negotiations between Britain and Argentina on 28 April 1980, the British delegation told the Argentines that they had to explain in depth all their proposals for the islanders; they were not to hide anything. If Argentina could not attain these standards, "they were bound to excite suspicion."[11] On the other hand, the idea of being on the side of the law was expressed numerous times: for example, "the legal justification for the Argentine claim to sovereignty is weak," contrary to Britain's "strong legal claims."[12]

At *Phases 2* and *3*, however, notable differences were found. The first is the frequency of the two arguments, with feelings of superiority outshining more explicit ideas of being a great power by 8.1 per cent to 1.2 per cent in *Phase 2*, and by 7.3 per cent to 2.8 per cent in *Phase 3*. The content of the two arguments also changed. The legal argument remained present but was relegated to a secondary role by the appearance of another argument based on military superiority. Immediately after the small garrison of twenty-two marines was forced to surrender in the Falklands and flown back to the UK, the marines stated that they "had not been impressed by the standard of the Argentine soldiers." The Governor of the Falklands, Rex Hunt, replied to Argentine soldiers who complained that he did not want to shake hands with them that "it was not civilised to carry out an invasion."[13]

Once the hostilities had begun, both arguments centred around the idea of Britain's military strength and effectiveness in comparison with that of Argentina. "We did not need to keep many naval units off the Falkland Islands to make life intolerable for the Argentines," explained John Nott, the British Defence Secretary to Haig.[14] Military strength is reinforced in the discourse by the fact that Britain was operating far from her bases compared to Argentina and operated with materials not always best suited for the missions they were tasked to accomplish, but despite that, the British still won and results obtained were much better than planned.[15] Despite heavy attrition rates being expected by the

British, the Harriers proved very efficient, thanks to superior radars and missiles,[16] causing Argentina to sustain heavy losses, even though Britain suffered some naval casualties.[17] The idea of military strength and effectiveness was then used to support the UK's position within a great power framework. As Parkinson wrote in his letter to the Conservatives: "The technical proficiency of our forces, which had been superb, have combined to make the Argentinians realise that Britain is not a nation to be trifled with."[18]

Specific British Roles on the International Stage: Enforcing International Law and Protecting Civilization

As Thatcher reported to Haig, German Chancellor Helmut Kohl was thankful that the British took the decision to send the fleet because "it was essential that somebody said 'enough.'"[19] Saying "enough" to the Argentine junta was an expression of the specific role that Thatcher claimed for Britain: enforcing and protecting international law, democracy, and civilization against authoritarianism.[20] While essentially correct (protecting international law and democratic principles were indeed important motivators), Gibran's implication that these ideas were in effect used to hide the real motives is a more problematic interpretation.

In fact, my analysis shows that the idea of having specific "roles" was present in 16.2 per cent of the dataset, with a distribution across actors and phases of the conflict that roughly mirrors the broader distribution of status-related ideas/perspectives. The view that Britain had specific "roles" to play on the world stage was an argument built around four ideas. The two most frequent ones were that Britain was the protector of international law (11.8 per cent) and the protector of democracy (8.2 per cent). The other two were far less frequent: being the world police (1.5 per cent) and being the mediator of international issues (0.5 per cent). These frequencies and the perspectives they reveal are reinforced by broader narratives surrounding each country's perceived roles on the international stage. For example, 5.3 per cent of documents displayed the idea of a conflict between democracy and dictatorship, while 9.4 per cent showed the idea of a struggle between the rule of law and authoritarianism. Britain saw itself as the protector of international law and democracy precisely because decision-makers saw these causes as being under threat. When coding for narratives of victimization, the islanders were most frequently evoked as victims of the Argentinian regime (7 per cent), a portrayal that coincided with the perceived necessity of safeguarding democracy and international laws and principles within some 3.7 per cent of documents.

The two main ideas composing this "international roles" paradigm are best expressed by Francis Pym (Defence Secretary) in his minute to Thatcher on 25 May 1982: "Our two central aims are to repossess the Islands and to determine their future, taking full account of the islanders. Other factors must not be allowed to override these aims."[21] The arguments, combined here as one, were not equally used throughout the conflict. At first, the idea of protecting democracy was more frequent (9.7 per cent of cases at *Phase 1*) than the idea of protecting international law (4.1 per cent), but both were used equally in *Phase 2* (5.5 per cent) during discussions on how to react to the invasion. Once the fleet had departed, the argument of protecting international law took dominance (19.8 per cent in *Phase 3* compared to 7.9 per cent for the idea of protecting democracy). Let's look in more detail at how these perceived roles changed throughout the conflict.

Before the Argentine invasion, the islanders often expressed to the British government their desire to remain under British administration. It is in this context that we find stronger evidence for ideas surrounding Britain as a protector of democracy. That the wish of the islanders was paramount was key in the British discourse and arose in many situations, strengthened as well by arguments that the islanders' choice should be respected according to the UN Charter principle of self-determination. In his instruction to Anthony Williams, the UK Ambassador in Buenos Aires, Micheal Fearn (head of the FCO South America department) said, "the Islanders have the right to determine their own future."[22] This idea reoccurs during the Anglo-Argentine talks in New York.[23] The idea of protecting the islanders' choice was found in the external documents, such as the letters Thatcher wrote to MPs who were asking her for details about her Falklands' policy, as well as in internal documents, such as the defence and overseas discussions where Thatcher summed up most of the debates by saying "they [the islanders] should be fully supported."[24]

When the decision to send the fleet had been made, the argument for protecting democracy receded and the protection of international law became the new paramount role for Britain, especially after UNSC Resolution 502 (demanding a complete withdrawal by Argentina from the Falklands) was taken on 3 April 1982, followed by Resolution 505 (calling on Britain and Argentina to agree to a ceasefire before negotiations could start) on 26 May. As Argentina refused to agree, it was necessary for someone to enforce the resolutions: a role that belonged to great power status and was thus claimed by Britain. In fact, in their discourse Britain linked its dispute with the Falklands to a much broader clash between rule of law and authoritarianism (found in 9.4 per cent of documents). As such, a key aspect of the conflict was "whether or not naked aggression should be allowed to succeed … [therefore the] aggressor must not be permitted to benefit from his aggression." This was framed as "more important than the fate of the islanders. If Argentina were not stopped, it would be a disastrous precedent for the world as a whole."[25]

Trying to manage the international system, thus enforcing international law, is a part of great power status, so it is not surprising to find this argument in numerous external documents. In all her various discussions with other heads of state and of government, Thatcher used the Falklands conflict as the prevailing expression of a wider conflict about the rule of law. She explained that if the British were to fail, many other smaller countries would potentially be in danger because nobody would come to their rescue. She also argued that there was a risk of seeing tensions increase everywhere should the UK fail. As she told Cameroon's President Ahidjo: "If this rule over boundaries, which was accepted the world over, were to be overturned, it would be a recipe for chaos. Many other countries and islands around the world would live in fear of invasion."[26] Thatcher told Reagan the very same thing when he tried to convince her to stop the fleet.[27] One finds again the return of a hero-villain discourse, this time through a status- and symbolic-oriented framing: Reagan was villainously putting international law in danger while Thatcher was heroically trying to protect those extant principles.

The Acknowledgment by British Officials That Status Is Relational

The above sections outlined evidence of British officials' conceptions of their great power status and what exactly this constituted. We see already that such status perceptions played a key argumentative role in the justification for going to war. However, British perceptions of status alone make up only part of the symbolic field of status perceptions in international relations. Another key question is whether British officials acknowledged the relational aspect of their nation's status, and how this relationality may have shaped the narratives and decision-making over status signalling.

Weighing the Options for Appropriate Status Signals

The theoretical framework of status-signalling is based on the idea that in order to gain status, claimants must demonstrate that they fulfil the roles and duties attached thereto. Doing so consists of sending the appropriate signals. In the case of the British reaction to the Argentine invasion of the Falklands, the question remains of whether British decision-makers believed that great power status could only be achieved through the recognition of such status by relevant others (i.e., peers) – which, in effect, would require Britain to convince them that it was entitled to great power status by acting like one. However, and quite surprisingly, the idea that status is gained through recognition by relevant others was only present in 0.5 per cent of documents. Despite this, explicit and implicit acknowledgment that status is gained by sending appropriate signals was evident within 13.8 per cent of documents. Such ideas were more frequent in *Phase 1* and came more from ministers and the public than from Thatcher (15.3 per cent and 23.1 per cent, respectively, compared to Thatcher's 13.7 per cent). That Thatcher had the lowest frequency provides further evidence in favour of the argument that the public wanted "blood."

So, how was the argument around sending appropriate status signals built, and was there any change over time? Before the invasion, the British government faced the difficult question of how to handle negotiations with Argentina. On one side were those arguing that negotiations were not satisfactory and were counterproductively sending the message of a Britain unwilling to negotiate: as Williams warned the FCO, "No negotiation can be convincingly confined to one day-and-a-half session a year, particularly when it has been sixteen years starting; we shall not, consequently, look reasonable in either the Argentine or world eyes."[28] On the other side were those who feared that negotiating with Argentina would be interpreted as a show of weakness. As Hunt reported to the FCO, "They [the Falkland islanders] have extreme concern that Britain appears to be abandoning its defence of British interests ... They feel that such a withdrawal will further weaken British sovereignty in this area in the eyes not only of Islanders but of the world."[29]

The issue of sending appropriate signals took a more dramatic turn when the decision to withdraw HMS *Endurance* was first made public, and once Davidoff's illegal landing in South Georgia became known. The withdrawal of HMS *Endurance* was questioned by the islanders and MPs to such extent that it became the main topic of discussion in parliament. Thatcher had to justify and explain the decision both in parliament and by personally writing to MPs. Her argument remained consistent, positing that HMS *Endurance* "has in any case, only a limited military capability," and that to withdraw it would have limited repercussions. She continued: "We remain determined to ensure that this integrity is preserved. We shall continue to maintain the Royal Marines Garrison at its present strength as a demonstration of our commitment."[30] To be fair, as an ice patrol ship armed only with a 20 mm gun, HMS *Endurance* had zero military value, although its withdrawal was misinterpreted by the junta.[31]

Once it was reported that Davidoff had illegally landed in South Georgia, the issue of sending appropriate signals was again addressed. Yet again, there were two opposing views. One favoured a diplomatic solution to avoid risking an escalation of tensions, as summarized by the FCO in its instructions to British diplomats in Buenos Aires: "This [the naval option to remove by Davidoff's group] could be a very dangerous impression to leave – even for a few days."[32] The other view, taken by the Falklands, favoured a more direct approach to the problem and warned that any lack of firmness could lead to further similar situations. "If our reaction is only placatory and is not firm and final, this time I [Rex Hunt, Governor of the Falklands] predict that more unopposed illegal landings will follow and probably next time somewhere in the Falklands."[33] The view of a necessity to react quickly and firmly to Davidoff's landing was shared by the FCO. Its response when asked what the reaction should be was, "a) the situation is intolerable, b) the remaining Argentines must be removed."[34]

Once the Falklands were invaded, divisions remained as to what type of signal should be sent: namely, how to find the right balance between negotiating and appearing firm. One cabinet position held that, although diplomatic means had to be pursued, sending the fleet was essential to strengthen the British position and show British resolve not to accept any fait accompli.[35] At the same time, there was the concern that Britain must appear willing to negotiate and not reject every peace proposal: "It would be important to avoid giving the impression of rejecting the United Nations Secretary General's proposals."[36] After all, since Britain claimed to be superior to Argentina due to its democratic values and standing as a mature country able to follow rules (ideas identified frequently in 10.7 per cent of documents), appearing stubborn would have been counterproductive.

Sending the Fleet to the Falklands and Great Power Status: Meeting the Criteria

From a symbolic point of view, sending most of the Royal Navy to the Falklands would have demonstrated British military expertise and supported its claims to great power status. The islands are 13,000 km from Britain and the nearest supply port and airport Britain could have used were at Ascension Island (6,000 km away from the Falklands), thus putting British strategic bombers out of range. Argentina operated at a distance of "only" 700 km from their bases. Winning a war in these conditions would certainly have restored British prestige in the world, proving that Britain had the capacity to project its hard power far away while demonstrating its financial capacity and resolve to do so. As the UK was in decline, a victory in the Falklands would be viewed as a great achievement and would surely make the world forget about the Suez fiasco, thus restoring the image of a strong and triumphant UK.

It seems that the British government was conscious of the Falklands crisis being the perfect opportunity to restore British credibility as a great power (before, during, and after the deployment of British forces), as evidenced by the relatively high use of the idea of sending appropriate status signals by both Thatcher and her ministers. Before the Argentine invasion, Cabinet Secretary Sir Robert Armstrong wrote a minute to Thatcher in which he argued that Britain's handling of the Falklands dispute would be decisive for British status and prestige: "A British readiness to surrender sovereignty over the Falkland Islands might have implications for the British position elsewhere in the world … What would the international reactions be if the UK showed itself willing to negotiate with the Argentines over sovereignty of the Falkland Islands? Would this be regarded as another example of enlightened statesmanship, or simply another sign that GB is on the skids?"[37] Interestingly, Thatcher annotated the note and underlined the last proposition, making clear that, for her, a retreat from the Falklands would be interpreted by the international community as a display of weakness.[38]

After the conflict, Thatcher repeatedly used the argument that victory in the Falklands re-established Britain as a great power, in both her speeches and her autobiography. "Our country has won a great victory and we are entitled to be proud ... Britain [still is] the nation that built an Empire and ruled a quarter of the world."[39] She also continued to praise Britain's special status compared to other countries: "The British are competent, courageous and resolute ... The lesson of the Falklands is that Britain has not changed, and that this nation still has those sterling qualities which shine through our history."

Use of the Past as a Reference Point to (In)validate Status Claims: An Argument Not So Evident in the Data

Another element in the theoretical framework of status-signalling holds that status assessments are made through reference points based in the past – not in the future, as Taliaferro[40] posits – and by undertaking a past-present comparison of all elements of great power status: capabilities, actions, roles, etc. To know whether British decision-makers were comparing their current situation to a previous one, the dataset was coded for evidence of references to timeframes of great power status. The results show that British decision-makers did not use any specific timeframe as a reference point, with evidence for present and future timeframes fully absent from the data, and past timeframes appearing in only 0.7 per cent of documents. Surprised by these results, I investigated any direct or indirect references to a "glorious past": such references were present in only 0.5 per cent of documents.

These low frequencies were even more surprising from a qualitative perspective. Where found, they were used by Falkland islanders in their discussions with FCO officials during the negotiation phase with Argentina. Islanders used "glorious past" references to show the FCO how determined they were to remain British. The clearest example can be found when a Mr Bennett, one of the councillors tasked with expressing the islanders' view to FCO officials, told Ridley that they would "rather pay the price of freedom. The Falkland Islands had supported Britain in World War II and had suffered casualties," implying that they would be ready to do the same again.[41]

Other minor references to the past were identified in the analogies used in support of military reaction (found in 3.4 per cent of documents), where reference to the Second World War and the successful British stand against Nazi Germany was used in interviews to the press, talks with other heads of state, and during internal governmental discussions.

Britain had faced dictators in the past and it decided to approach the situation in the South Atlantic in more or less the same way.

Despite the low frequency of direct references to the past as reference points of great power status, the analysis did identify a nostalgic narrative in some 4.4 per cent of documents. While rather minor in frequency, this narrative still merits analyzing in combination with status considerations, which we will turn to in the next section.

War for and because of Status: A Conclusion

Overall, the evidence makes clear that there was an important status and symbolic component to the British decision to go to war over the Falklands. This component took on several shapes: descriptions of the geopolitical situation, framing of the adversary, understandings of the conflict, and the incorporation of this particular conflict into a much bigger one (i.e., against authoritarianism and international lawlessness). We find that British decision-makers were clearly aware of the status perceptions surrounding the conflict as they carefully debated their options for sending appropriate status signals. How exactly they conceived of the relationality of the status-recognition process, however, is less clear. The higher frequency of findings within external and public documents could indicate that there were at least two levels acknowledged by decision-makers in the status-recognition process: domestic recognition and relevant-others recognition. Indeed, the possibility that ideas surrounding status were used more frequently in public documents to reassure public opinion that British status was being appropriately managed – thus producing a situation of a status-orientated DTW – should not be ruled out.

The analysis also uncovered some positive evidence regarding status considerations as arguments for a military response and several indirect references to the past (e.g., Munich and the Second World War), although references were too infrequent to draw reliable conclusions. Based on these findings, we may conclude that status and symbolic aspects played an overall important, although not necessarily dominant, role in the British decision to go to war. As the next section will outline, status and symbolic considerations maintained important affinities with the framework of nostalgic virility, thus highlighting a number of novel explanatory perspectives on the British reaction to the crisis.

Nostalgic Virility: Showing Them That the Lion Can Roar Again

To deepen our understanding of the role of status perceptions in the British decision, we will now turn to an evaluation of nostalgic virility – a closely related element to status perceptions which, among other things, incorporates feelings and identifications of strength and virility into one's self-understanding and view of their status.

Nostalgic Virility, a Surprisingly Dominant Framework: What the Numbers Reveal

Evidence directly related to the framework of nostalgic virility was found in a surprising 40.7 per cent of documents analysed, with the frequency increasing to some 58.4 per cent when accounting for indirect yet related findings. These initial numbers point to nostalgic virility as a dominant framework for understanding the British decision to go to war. At the quantitative level, the stark difference in frequency between evidence for nostalgic virility and the other frameworks tested above demonstrates that the former likely has significant explanatory power, at least in the context of the Falklands crisis. This must be verified through the qualitative reading of the data. However, before turning to this let us first consider what the frequency distributions of evidence for nostalgic virility might tell us.

The frequency of evidence for ideas and perceptions related to nostalgic virility was on the rise all throughout the conflict (26.9 per cent at *Phase 1*, 29.2 per cent at *Phase 2*, and 59.3 per cent at *Phase 3*). When breaking the results down by actor, we find that this framework flourishes in documents issued by Thatcher (72.6 per cent) and the public (59.3 per cent, including for MPs, the press, islanders, mainland domestic opinion, etc.), while ministers score much lower yet still significantly (29.2 per cent). An initial question posed by these results is whether or not Thatcher's frequency is a genuine representation of her own feelings and perceptions or whether she employed such rhetoric and argumentative logics in the knowledge that the public would be receptive thereto? To gain an initial insight into this, I cross-tabulated the results for *Thatcher* by document type. The findings were unexpected: the framework is present in high proportions in all types of documents, although significantly higher in public and external ones (57.8 per cent

Table 6.3 Frequency of the four theoretical frameworks in the Falklands case

Framework	*Frequency (per cent)*
Realist-strategic	3.6
Diversionary theory of war	13.6
Status and symbolic	17.4
Nostalgic virility	40.7

in internal documents, 93.8 per cent in external documents, 93.5 per cent in public documents).

An initial interpretation of these findings suggests that the British public, whatever the reason being, was very sensible and responsive to the idea of virility. But what did virility mean or represent to these different actors? Was it based on the same ideas or logics? And were there differences in Thatcher's perspectives when communicating publicly or externally compared to her internal communications and deliberations? Based on the nature of the framework and the findings we have uncovered thus far, we should, theoretically speaking, find that in *Phase 1* the framework is built around the idea of sticking to the position that islanders' wishes are paramount. In *Phase 2*, the prominent argument should shift and be based on feelings of outrage and humiliation. Finally, in *Phase 3* the framework should be centred on the need to punish Argentina and not appear weak. These questions and preliminary postulates are investigated below through a qualitative reading of the data.

From Feelings of Loneliness and Humiliation to Feelings of Joy: The Evolution of British Emotions throughout the Crisis

Emotions are often forgotten or omitted from research and analysis in the field of IR. In the case of the Falklands crisis, doing so would force one to ignore a large – and important – part of the story. The use or expression of emotions were found in 26.6 per cent of the dataset, but only in *Phases 2* (19.8 per cent) and *3* (29.9 per cent). However, based on a qualitative reading these were rarely found to be explicit, meaning that these percentages should be taken with a grain of salt. The emotions identified were, expectedly, diverse – something that Helen Parr has suggested is the norm in these types of situations.[42] However, among the most prominent and substantively significant of emotions expressed during the crisis was the feeling of humiliation, which was found to be shared by an array of British actors including Thatcher, cabinet ministers, MPs, the military, and diplomats.

Feelings of humiliation were identified during the Davidoff incident when it became clearer that the Argentines had no intention of settling the dispute peacefully, contrary to what they told British diplomats and claimed publicly. For British diplomats, the feelings of humiliation were initially based on having been easily manipulated by dishonest interlocutors: "I fear that events over night enforce the growing impression that Costa Mendez [the Argentine Foreign Minister] has been less than honest with me and that, in fact, the Argentines have been playing us along."[43] These feelings were accentuated during talks with US officials, as it appeared in the eyes of the US that responsibility for the invasion should be shared equally between Argentina and Britain. This view was unacceptable for Britain, one of the world's oldest democracies, as it was being placed on the same level as Argentina – a country living under the rule of a junta with a terrible human rights record. As noted by Carrington: "I was upset by [the] attitude [of the US] … It was rather odd to see ourselves put on the same footing as the Argentines."[44]

News of the Falklands invasion led Conservative MPs to a boiling point. On 3 April, the first post-invasion "1922 Committee" meeting was held.[45] Gow, Thatcher's private secretary, took notes on this rather heated meeting, noting that several members qualified the situation as a "miserable day" while others asked "Why did we sell arms to Argentina?" and "Why the hell were we negotiating?" Others argued that there was no chance of reconquering the Falklands because of arms sold to the Argentines (especially French Exocet missiles) which would "wipe us out" and "sink the fleet." On the other side, there were a few MPs calling for a strong response: "We must be tough now."[46]

Not to be outdone, the military personnel present on HMS *Endurance* described it as a "humiliating day,"[47] while British diplomats spoke grimly about an "affront."[48] The feeling of humiliation was said to be accentuated when pictures of Royal Marines surrendering (made to lie face down on the ground) were made public by Argentina.[49]

Humiliation was not the only "negative" emotion present during the conflict. Some 5.3 per cent of documents displayed markers of a feeling of "being alone" against all others, which was identified in all

Figure 6.1 Surrender of Royal marines in Port Stanley.

three phases of the conflict (8.3 per cent, 3.5 per cent, and 4 per cent, respectively) and first evidenced during pre-invasion negotiations with Argentina. Even this early on Britain complained that "the majority of the UN membership including the US and most of our Western friends are openly sceptical about the legitimacy of our position in the Falklands."[50] In fact, the UK's isolation on this front was not new. It was already clear when looking at the results of the 1976 UN Resolution vote, which showed "102 in favour [of the pro-Argentine resolution], 1 against (ourselves) [Britain] and 32 abstentions."[51] Through British eyes the situation was made even worse as fears grew that the US would step in and force them to adopt a settlement over the Davidoff incident which would go against their interests. "We fear that we would end up with the Americans also exerting pressure on us to reach a solution which was politically unacceptable to us."[52]

These fears increased when the fleet was on its way to the Islands, as Haig made clear during his shuttle between Buenos Aires and London that, for the US, "there cannot be another Suez,"[53] and that the UK's other allies would be "reluctant to take up actions damaging to their own economic interests."[54] It was vital for Britain to have the US and other traditional allies on-side, hence the hero-victim-villain narrative (see chapter 5).

However, in the time between 25 April 1982 and the Argentine surrender on 14 June, the British mood changed entirely. On the occasion of the reconquest of South Georgia, Thatcher asked journalists to "rejoice at the news and congratulate our forces and marines."[55] The dataset stops on this date, at the end of the conflict, although signs of positive emotions and happiness from the British government would surely have been high – a complete reversal after weeks of fear and humiliation.

Reporting of emotions is, of course, generally absent from key internal government documents, such as cabinet meeting transcripts. As such, determining the degree to which emotions played a role in the decision to go to war faces several difficulties. Nonetheless, based on the evidence identified in this analysis it is clear that emotions were present and likely to have played a crucial role in the overall decision to go to war. Although one cannot assert that they were *the* reason or deciding factor, their role cannot be neglected, and their significance can be further highlighted by considering them in conjunction with the broader set of perceptions and motivations identified in this analysis. Indeed, our understanding of the role of emotions – namely humiliation and fear – in the decision to go to war is strengthened when considering the dynamics of nostalgic virility as (in part) a response to such emotional states.

A Conflict over and of Virilities: "Real" versus "Fake" Virilities among Thatcher, Galtieri, and the Advocates of Peace

As a counter to the feelings and perceptions of humiliation, identification and displays of strength and virility played an important role in the British discourse and deliberations, helping to construct Britain's own self-identity in relation to Argentina and other states (and thus further elaborating upon the relationality of status considerations). This section will demonstrate how Britain's "true" virility was constructed and perceived in relation to the "fake" virilities of dictators such as Galtieri.

In her article on British and Argentinian war speeches, Andrea Roxana Bellot argues that British public discourse had a distinctly patriarchal tone, with heavy use of masculine-toned words such as "honour," "kingdom," "blood," "enemy," "war," "arms," etc.[56] To test whether this patriarchal tone was consistent throughout the conflict or merely present at times, both conscious and unconscious use of such words and tones comprising a virile discourse were coded for investigation. Overall, such discourses were present in 13.8 per cent of documents, with an especially high frequency in documents aimed at the public (31.3 per cent). They also had a considerable presence within documents sent to other countries (18.8 per cent). One key question was whether Thatcher – with her title as the "Iron Lady" – used this type of discourse? Analysis shows that she did in fact employ viral discourses within some 19.7 per cent of documents, with an especially high proportion in her public discourses (41.9 per cent) and in those to other heads of state and of government (18.8 per cent). With these initial findings, it is clear that Thatcher frequently used virile components within her discourses. However, do these frequencies alone qualify her as a nostalgic-virile individual?

Nostalgic-virile individuals are those who emphasize strength either on a personal level (e.g., their body) or at a symbolic level (e.g., their country or military's strength). By the same logic, they also emphasize the need to be fit to lead, either at a personal level (e.g., being in good health) and/or a symbolic level (e.g., their country having moral superiority). For these reasons, they cannot bear the idea of "losing face" and fear any potential humiliation. Should this occur, it must be atoned as soon as possible, meaning they often prefer short-term solutions. These individuals like to appear firm and refuse to give up or make "U-turns." They express a significant need for high-level power with a low-level of complexity, meaning they often proceed to a clear-cut or dichotomous understanding of situations. Finally, they are nostalgic for a glorious past, with such nostalgia expressing itself through positive references to previous eras and through a desire to reproduce historic, strong behaviour.

Overall, most of the elements recapitulated above were found in the dataset, including an emphasis on strength (7.5 per cent), standing firm (18.6 per cent), and being brave (8.7 per cent). These aspects are far more evident in Thatcher's documents (12 per cent, 15.2 per cent, and 20.5 per cent, respectively) than in ministerial ones (5.8 per cent, 11.7 per cent, and 2.2 per cent, respectively). In addition, evidence of low-level complexity is found in 4.5 per cent of documents, but is effectively concentrated in Thatcher's documents alone (14.5 per cent compared to 1.5 per cent for ministers). This is not surprising as Dyson, in his study of Blair's foreign policy, notes that Thatcher was one of the British prime ministers with the highest low-level of complexity score.[57] Finally, evidence of nostalgia (addressed in the next section) was also found at a fairly low frequency (5.1 per cent), and once again concentrated in Thatcher's discourses as compared to her ministers (7.7 per cent compared to 2.9 per cent). An initial impression of these frequencies would suggest that Thatcher indeed fits squarely into the nostalgic-virile category, whereas her cabinet ministers tended not to.

So, how are the various aspects of nostalgic virility evidenced in Thatcher's discourses, and are they consistent with the theory's general predictions? From a qualitative reading, the first trend that stands out, regardless of the phase of the conflict, is a discursive combination of firmness and bravery. The best example of this combination comes in the wake of the HMS *Endurance* withdrawal, which had been a major topic of enquiry from MPs and others. In her letters of response, Thatcher linked two ideas: that such a decision was very difficult and sometimes heart-breaking to make, but that it was necessary because of superior interests; and that, despite the withdrawal of the ship, Britain remained as determined as ever to keep the Falklands British.[58]

Since 1981, when the topic was first discussed, Thatcher had stuck to her decision. Although the above quote was from 1982, one can find similar framing in documents from 1981.[59] She stood firm on her desire to support the islanders' wishes, which were continuously described as "paramount." Once the invasion became public knowledge, tricky decisions had to be made, such as sending the fleet, which could potentially lead to heavy casualties: "Difficult decisions would need to be taken if the Islands were occupied by Argentina. A British response would be necessary."[60] And, no matter what the costs, "we shall persevere until freedom and democracy are restored … The costs now, however high, must be set against the costs we would one day have to pay if this principle [rule of law] went by default."[61]

The second aspect of "conventional" virility identified was that of strength, expressed at both the personal and symbolic level. It first appears at the news of the heroic British defence of Port Stanley: "They inflicted casualties, but those defending Government House suffered none."[62] British military strength and bravery were repeatedly highlighted during the conflict, especially after the first loss of ships when focus was placed on the determination the military had shown in their search for the wounded: "[The] greatest bravery of the helicopter pilots on going to get the sailors who were wounded out."[63] Strength, or appearing strong, often go alongside the rejection of weakness – something that virile individuals abhor. Thatcher furiously stood up against such accusations: "any suggestion that our capacity to respond to the crisis has been weakened by a run-down in the strength of the navy is nonsense … The fact that the fleet has been able to put to sea, prepared for action, in four days, reflects great credit on the Royal Navy."[64] She wanted to portray Britain as strong, its capabilities increased by new ships and submarines that would be commissioned later.[65] The ultimate proof of British military strength? The fact that Argentina became willing to open discussions as the fleet steamed south. It was this British strength that achieved diplomatic progress, something the US negotiation shuttle had been unable to do. As Thatcher told Reagan, "The truth was that it was only our military means which had produced diplomatic results."[66] The idea of being fit to lead was found under various auspices that will be analysed in depth in the next section. Here, I will just posit some broad thoughts. Explicitly, the argument was found under its symbolic component: Britain was "fit to lead" because it is a democracy and should therefore provide an example to the world. "It should be for democratic nations to set the world as an example."[67] Another aspect of this argument is found under the framing of protecting others against evil (3.4 per cent) and punishing morally deviant actors (18.8 per cent).

In addition to these first elements of nostalgic virility, my analysis has shown that there were also indications of low levels of complexity (14.5 per cent of documents), as argued by Dyson.[68] He makes the argument that individuals have a low-level of complexity when they resort to an absolutist way of thinking and frame situations in clear-cut or dichotomous terms. These two markers were found within 10.3 per cent and 13.7 per cent of Thatcher's documents, respectively. There are several clear-cut framings in her discourses. She described the two opposing camps (Britain and Argentina) in simplistic terms: "We are right, they're wrong. We are the victims, they are the aggressors,"[69] because it is a "dictatorship versus democracy"[70] type struggle in which democracy is in the "right" camp.

She showed absolutist thinking, as well, in the way that she described the British response – "We have to recover these Islands"[71] – meaning it was a do-or-die situation in which there were only two options: victory or defeat. Refusal to even think about the possibility of surrendering the islands, which would have been considered totally unacceptable, marks her belief in the capacity to control events and avoid losing face at all costs.[72] Indeed, we find similar evidence of Thatcher's low level of complexity when she makes references to immutable natural laws – for example, positing that failing to stop a dictatorship early on leads necessarily to global unrest in the future.[73] According to Dyson, individuals with a low level of complexity are less likely to favour compromise. This is true of Thatcher, evidenced by markers in the dataset such as her refusal of any proposition of third-party mediation – whether from the US, the UN, Peru, Mexico, or Colombia – as long as Argentina had not withdrawn from the Islands (especially once the Royal Navy was already on its way to the Falklands).[74]

On the other side of the spectrum were Thatcher's physical and symbolic opponents: "wimps" and "sissies." They were described as such in British documents because they did not have the character traits associated with nostalgic virility. The first to fall into this category were members of the Argentine government. The Minister of Foreign Affairs, Costa Mendez, was described as indecisive and "without influence." The member of the junta on whom the British concentrated their criticisms was, of course, Galtieri. He was described by British cabinet ministers as "alcoholic and apparently incapable of rational thought,"[75] and unable to take difficult decisions. "They haven't withdrawn them (Argentine troops) for the last seven weeks. I always thought it would be difficult for a dictator to withdraw."[76] Added to these traits, Thatcher framed him as mad, cruel, and lacking a sense of leadership and reality. The foundation for this description is that he "had said over the weekend that he was prepared to sacrifice 40,000 lives" to fight against the British,[77] instead of acknowledging defeat, cooperating with UN resolutions, and starting negotiations with Britain.[78] On other occasions, Thatcher repeated to various interlocutors who were pressing her to stop British military operations that she could not do so because the Argentines, and thus Galtieri, could not be trusted: "There's absolutely no guarantee that at the end of the period, if they haven't got what they want, they won't just invade again."[79]

Once the news of the invasion became known, there were calls for the government to take direct and forceful actions. However, as the fleet steamed south and as the conflict gained in intensity (and casualties), the cabinet (especially Thatcher) had to face calls to order British troops to halt their advance. At an international level, these came from the Organization of American States (OAS), which was slow to support Britain. In her discussions with Haig in April 1982, Thatcher expressed dissatisfaction about their slow response. "I don't know of any clause in the OAS which applauds a state for taking territory of any other state by force," she complained, implying they lacked the resolve and courage to such an extent that "if it [the OAS] doesn't condemn this, they're all liable to have aggression by one state or another on their territory."[80]

At the domestic level, she was asked in a letter from Michael Foot, leader of the Labour Party, to stop the fighting and open discussions with Argentina. In her response, Thatcher indirectly portrayed Foot as weak, lacking resolve and bravery, and being a traitor to the military, as "It would be unthinkable to negotiate about the future of the Islands as if everything were still as it had been before. That would be a betrayal of those whom we have called upon to make such great sacrifices," especially after casualties in the British ranks.[81] This criticism was later extended during her interviews with the British press to anyone wanting Britain to stop military action: "All right, hand something to the Argentines. Under these circumstances to an invader, to an aggressor, and a military dictator, that would not be magnanimity. It would be treachery or betrayal of our own people."[82]

If the opposition to, and even the fight about, virility was essentially found between Thatcher and her adversaries, the qualitative analysis of the dataset also seems to indicate that there was debate on the subject between Thatcher and some of her key allies. This opposition to the notion of virility is first found, once again, when the future of the Falklands was discussed within the British government before the invasion took place. Thatcher relentlessly opposed her own Foreign Secretary from as early as 1979 when he suggested that transferring sovereignty to Argentina was the most rational decision because it was less costly.[83] Carrington favoured the lease-back option and, aware that Thatcher was against such an idea, instructed FCO staff to make the lease-back option appear under a more favourable light in any memos they sent her. What seemed to be at stake in these redrafts was not just the question of presenting options, but also of understanding the symbolic images of strength and other virile values surrounding them. The following passage highlights, indirectly, how these issues were important to her: "The Secretary of State would like the minute to the Prime Minister … to be recast in a more positive light … Lord Carrington considers that it launches too quickly into the leaseback option, which is presented as something, forced upon us rather than, an opening which we should try to exploit … The Secretary of State does not wish the minute to sound defeatist."[84] Thatcher's answer remained the same: *I could not possibly agree; the islanders' wishes remain paramount.* She handwrote these comments in a document she received from and sent back to Carrington,[85] but also on briefing documents forwarded to other members of government and the administration,[86] as well as on instructions to the governor of the Falklands.[87]

This opposition of political views took an explicitly virile turn on at least two occasions. On 12 October 1979, Carrington sent Thatcher a minute prepared for the Defence and Overseas Committee. It was heavily (and possibly angrily) annotated by Thatcher. On top of the document she wrote, *I don't like this paper.*[88] When Carrington wrote *Argentina has a long-standing and active claim to the island*, her answer was: *What?* (underlined in text).[89] Finally, when Carrington wrote *Argentina's claim is not just a matter of law but of national honour and machismo*, thus suggesting it would be wise to follow the lease-back option, Thatcher's answer clearly indicates that it was an issue of virility for Britain, too: *According to the Foreign Secretary our national honour doesn't matter!?*[90] Such opposition was indirectly acknowledged in Carrington's resignation letter on 6 April 1982, which was – interestingly – immediately followed by praise for Thatcher's virility-related traits.[91]

The second occasion at which the opposition between Thatcher and some of her ministers took a virile turn has to be considered in context. The discussions were not officially recorded in the minutes but can be found in Armstrong's private and personal notes. Apparently, during the Defence and Overseas Committee of 7 November 1980, Ridley, the junior minister in charge of the Falklands issue, claimed that Britain did not have the military capability to protect the Falklands against Argentina. Thatcher's answer was, apparently, very passionate and virile: "We could bomb Buenos Aires if nothing else."[92]

Punishing Argentina: A "Strict Father" Narrative with Many Justifications

After having developed self-perceptions and feelings of superiority in relation to Argentina, we find that British officials effected a discursive turn towards moral narratives of *punishment*. For example, evidence for George Lakoff's "strict father" moral narrative was present in 16.2 per cent of documents across the dataset (increasing from only 3.4 per cent to 15.1 per cent after Davidoff's landing, and again to 27.1 per cent after the fleet departed). Among British officials, Thatcher used this moral narrative most often (39.3 per cent of documents compared to 9.5 per cent for ministers and 12.8 per cent for MPs and the press). This moral narrative was also much more evident in public (40.3 per cent) and external documents (31.9 per cent) than in internal documents (6.2 per cent). To compare with the "inverse" of the strict father narrative, evidence for Lakoff's "nurturant parent" narrative was also investigated. Analysis reveals that this moral narrative was present only at a much lower frequency (1.2 per cent) and only prior the Davidoff incident.

Before unpacking the substance behind these frequencies through a qualitative reading, let's first consider what these initial numbers might suggest. The first indication that such high frequencies of the "strict father" narrative suggests is that, based on respect for international law, punishing Argentina was viewed as a moral necessity due to the country's deviant behaviour. The second is that, as the framework is more evident in public and external documents, there was a need to punish Argentina to please public opinion (in which case we should see the revival of a specific DTW perspective). Alongside these two preliminary observations is a third that requires testing. Status considerations were strongly present in deliberations, meaning that the argument for punishing Argentina in order to match and uphold Britain's great power status should not be forgotten as a potential element of the strict father narrative. These three observations will be tested below through the qualitative analysis of the data.

Moral Superiority and Virility: Punishing Morally and Behaviourally Deviant Actors

In the previous subsection, we also saw that Thatcher used ideas of being "fit to lead" in a symbolic way, holding that "it should be for democratic Nations to set the World an example."[93] How would Britain set such an example?

First and foremost, by being a mature and disciplined country. As agreed by the British government in Anglo-American talks during Haig's shuttle, "We are a responsible power and the Argentines are not and this imposed certain obligations, and indeed problems, for us in the eyes of the world."[94] Being a responsible, mature, and disciplined country meant that Britain should follow and play by the rules, namely international ones. That is why the British, especially Thatcher, kept reiterating to other heads of state and government that Britain was in favour of a negotiated settlement with Argentina – despite evidence outlined above that Thatcher was not in fact interested in negotiating. "We must make a supreme effort to prevent a major military clash," Thatcher wrote to Reagan.[95]

The fact that Britain was at first willing to negotiate and that the fleet was sent primarily as a back-up for any negotiations was also central to the argument she used in her replies to questions from MPs and others: "They [the islanders] must be freed from occupation. Naturally, we would prefer to achieve that peacefully, and we continue to search for a negotiated settlement."[96] Being mature and disciplined also meant providing specific rules of engagement, with the objective of fewer casualties on both sides. This was initiated as early as the Argentine invasion when "it had been agreed not to fight in Port Stanley because of possible civilian casualties."[97] Once the fleet left Britain, the Royal Navy received instructions that Argentine units should only be engaged if they were located in, or steaming to, an exclusion zone located around the Falklands, thus upholding international law. "We had been careful, for example in our announcement of an exclusion zone, to observe international law scrupulously."[98] Targeting Argentine aircrafts was also subject to specific rules of engagement.[99] And, of course, it was made clear publicly that it was "inconceivable that Britain would use [nuclear weapons] in confrontation with Argentina."[100]

The contrast to this narrative portrayal of Britain as strong and disciplined was the portrayal of Argentina as guilty of immoral and deviant behaviours that should be severely punished (evidenced in some 7.3 per cent of documents). In British discourses, Argentina was seen as guilty of exhibiting deviant behaviour in several ways. It was accused of being dishonest and (ironically) of having never been in favour of real negotiations, to such an extent that "one is bound to ask whether the junta have ever wanted to seek a peaceful solution."[101] The Argentines were seen as willing to use negotiations to gain time by refusing peace proposal after peace proposal. Thatcher used this argumentative line in all types of documents. When the President of Colombia asked her to keep negotiating she replied, "I am sure you will recognize that in the seven weeks of negotiations in which we have engaged in good faith," making sure he understood that this was not the case for the Argentines, they "refus[ed] to match our constructive approach."[102] Instead, Argentina was accused of using the negotiations to gain time to strengthen their position, a behaviour that could not go unanswered: "The government could not allow Argentina to continue with the idea of prolonging the negotiations in order to weaken the political, economic and military pressure which the United Kingdom could bring to bear."[103]

In Thatcher's narrative, Argentina is a morally deviant state (because of its political regime) that ignores reason and responds only to strength. Because of this, Argentina should be "guided" to find a path to the negotiating table. "It [is] essentially an issue of dictatorship versus democracy. Galtieri wanted to be able to claim victory by force of arms. The question now was whether he could be diverted by language he understood."[104] To "help" the junta, sending the fleet was necessary as it was the ultimate guarantee to show that being morally superior does not mean being weak, and Thatcher made crystal clear that Britain might be a democracy and might be mature and disciplined, but that did not mean it would not react forcefully to any violation of international laws and principles. The moral necessity to resort to military force was strengthened from 29 April 1982 onward. Two days earlier, Haig had sent peace proposals to Argentina with a twenty-four-hour deadline. Two days later, in the absence of any Argentine reaction, Thatcher told Reagan that "this deadline has passed more than 24h away … In the cabinet's view, the proposal must be now regarded as having been rejected by the Argentines, who have ignored the deadline."[105]

The various examples identified in the data of logics or narratives seeking to punish morally deviant behaviours show that the British government, especially Thatcher, tended to define policies as rules that operate through or incorporate a biopolitical dimension. Such policies articulate governmental rationalities (and sometimes narratives) with the stated aim of providing for the health and welfare of populations.[106] As Argentina disrespected and violated norms of the international community, it represented a danger or a disease that had to be cured by an appropriate treatment (i.e., war) in order to safeguard the health of the international system.

Back to the Future! Yes, but Why? Nostalgic Virility or Pleasing Domestic Opinion?

The theoretical framework of nostalgic virility holds that nostalgia has an important role in the decision to go to war as it influences virile individuals in favour of military action. In these instances, nostalgic sentiments make individuals feel that they should perform the same type of actions and send the same type of signals as their predecessors. Virility is not displaced by such nostalgia, as it serves rather to make nostalgic individuals reject any situation that might show modern-day weaknesses (real or perceived). It makes them envious of the past, of times when their country was strong and highly considered by other actors on the international stage. The answer to any challenge can thus be found in the past, which provides glorious examples.

As summarized above, no specific time periods of great power status were preferred or evoked by British officials to any considerable degree, while overall, nostalgic discourse was present in only 4.4 per cent of documents. The question remains, despite this relatively low frequency, of whether the nostalgic narratives had any impact on the decision to go to war? Before turning to a qualitative reading of these narratives, let's first consider their distribution. Nostalgia was mostly present after the fleet departed (5.1 per cent of documents), although evidence was also found before the Argentine invasion occurred (4.1 per cent). Among actors, the most sensible to these narratives were Thatcher (7.7 per cent) and the public (MPs and the press, 8.6 per cent), and interestingly, the results become considerably more frequent when also accounting for the islanders (evidence rises to 10.3 per cent of documents). Furthermore, nostalgic narratives are found mostly in documents sent to other countries and aimed at the public. These results indicate the possibility that the need to intervene militarily to punish Argentina was motivated partly by nostalgia, and that this was, in part, in order to appeal to the nostalgic currents within domestic opinion.

Thatcher constructed a nostalgic discourse around the idea that Britain was a strong country that had set up throughout history a certain number of moral behavioural rules on the international stage, and that these rules were guidelines which Britain would act upon. If Britain could not do so, then it would not be Britain: "We would not be the country we are unless we were prepared to fight for these things. To fight for freedom and justice and liberty ... Just as our father and forefathers fought for us."[107] This argument was made during each of her interventions in the media,[108] as well as during her conversations with other heads of state. For example, she told Helmut Schmidt, then German Chancellor, that if Argentina forgot who Britain was, they would soon find out: "The Argentine regime will know what to expect. Britain does not appease dictators."[109] The issue at stake for Britain was whether it could still stand up against dictators: if Britain could not act as she had in similar situations in the past, then that would send a negative signal about both British capabilities and identity, potentially leading to a loss of status.

Using a Second World War analogy, Thatcher took on the persona of Winston Churchill and argued that to avoid any worsening of the situation, dictators must be stopped as early as possible or there would be a repetition of the 1938 and 1940 situations. From her speeches, it is obvious that Thatcher did not plan to wait any longer, as she justified in her ITN interview on 5 April 1982: "If you stop a dictator there are always risks but there are greater risks in not stopping a dictator. My generation learned that a long time ago."[110] This Second World War analogy is consistent with other findings on Thatcher's use of the Munich analogy.[111] Thatcher portrayed the two situations as similar and was determined that, this time, Britain would not step down.

Continuing the Second World War analogy, Churchill took the risky decision to continue the fight against the Nazis despite a succession of crushing defeats prior to 1942. In Thatcher's case, the risky decision was to send the entire Royal Navy to the South Atlantic.[112] This was risky because of the operational difficulties of having no supply port in the area of conflict and relying on aircrafts not designed for air superiority to protect the fleet. In their military study of the conflict, Hastings and Jenkins make reference to the gamble: "As the nation's only throw, a huge commitment of national prestige and residual power, the dangers were enormous."[113] If a carrier were to be lost, the mission would become impossible:[114] "After the failure of the Royal Navy's attempt to gain air superiority before the arrival of the amphibious force, the government and the service chiefs committed themselves to an even greater gamble, that of sending the landing forces regardless."[115] As Gibran writes, "A price had to be paid for the restoration of Britain's image and credibility and that price involved going to war with Argentina."[116] It was vital to punish Argentina. Not only for what it had done from a material perspective (invading a democracy), but also for what its action meant on a symbolic level (challenging international law and principles), all of which culminated in an unacceptable challenge to Britain's virile identity and past.

Takeaways from the Nostalgic-Virile Narrative

Based on both the quantitative and qualitative analysis of the dataset, we may affirm the importance of the nostalgic virility framework in the British decision to go to war. Critical components of the framework were found as attention was drawn to the ideas of strength, bravery, and being fit to lead. In the case of nostalgia, this element appeared to be less important than other aspects surrounding virility. In both cases, however, links to broader status considerations can also be found, thus providing evidence of nostalgic-virile understandings of great power status. The idea of punishing Argentina can also be linked to that of pacifying domestic opinion, suggesting the possible existence of a virile element to the DTW – namely, demonstrating to the public that Thatcher still had what it took to lead Britain. On the whole, while it can be argued that nostalgic virility played a key role in the British decision to go to war, it was not possible to find conclusive evidence in favour of either of these frameworks as representing the primary determining element or argumentative logic used for making the decision to go to war. Nonetheless, they were clearly important and very present within the ecosystem of influences and deliberations leading up to Britain's reaction.

General Conclusion: Which Framework Best Explains the British Response?

I began my analysis of British military reactions to the Falklands crisis by providing a priori evidence in support of each theoretical framework being tested. The frequencies and qualitative analyses of the dataset have validated some and not others, often with surprising findings.

The mainstream frameworks – realist-strategic and DTW approaches – obtained mixed results. It is clear that the realist-strategic framework did not play any important role in the decision to go to war and was mainly used to build an argument to gain support from the US at a critical moment for Britain. Results for the DTW approach were far more nuanced. If, in order to consider the framework as relevant, one would expect to find clear evidence of a risk of losing power in the absence of a military response, then this was not the case and the DTW had little explanatory power. While there were some political considerations from MPs and cabinet ministers, there was little to no evidence of such considerations from Thatcher, except in her discussions with the US, which pointed toward the idea of losing power as a strategic narrative used to secure their support. What is much more ambiguous is if the decision to go to war was motivated by a recognition that the British public wanted "blood." I found some evidence to support this view, but from the available data it is not possible to conclusively assert it nor rule it out. In any case, the fact that British public opinion might have expected "blood" in fact reinforces the nostalgic virility framework. It shows the potential importance, in some instances, for leaders to be nostalgically virile in order to demonstrate their concern for the nation's status and their resolve to not let challenges go unanswered: in essence, that they are willing to do everything possible to equate with their country's glorious events of the past.

The status and nostalgic virility frameworks are built around the ideas of sending appropriate status signals and, in conjunction, of upholding the symbolic (or concrete) perception of nostalgic virility. On the balance of evidence, these complementary frameworks seem to be the most robust in explaining why the British decided to go to war against Argentina to reconquer the Falklands. Key components of the status framework, such as the need to send appropriate signals and the view that Britain has particular roles and duties, were identified as important argumentative logics in the decision to go to war. As well, key elements of virility were found along with some (albeit far less) evidence for nostalgia, both of which played symbiotic roles with broader status arguments and perceptions, resulting in what appears to be a very strong influence on British decision-making (notably for Thatcher).

The overall picture obtained through the quantitative and qualitative analyses of British declassified documents is that traditional arguments given in the literature to explain the British decision to re-conquer the Falklands are generally inaccurate and find minimal support (if any) in the available data. Instead, what seemed to be the primary trigger for the British reaction was the virile nostalgic interpretation of Britain's great power status. Letting the Argentines' aggression go unpunished was seen as simply unacceptable from both a status point of view (i.e., losing any chance of reclaiming great power status) and a nostalgic virile point of view (i.e., not following in the footsteps of past British governments' opposition to invaders, accepting a public humiliation, and showing further weakness). While realist-strategic and DTW frameworks both find some evidence or justification within the data and contribute important elements to the story, we find that they nevertheless fall short in explaining the British reaction: a decision to go to war which cannot be fully understood without consideration of the role of status and nostalgic virility.

7

The US Invasion of Grenada, 1983 (I)

Testing Mainstream Theories

On 25 October 1983, an impressive task force centred around and under the command of US forces stormed the Caribbean country of Grenada. The task force included an aircraft carrier strike group, an amphibious assault ship, marine units, ranger units, the famous 82nd Airborne Division, Navy SEAL commandos, and allied forces from other Caribbean states – all to invade the small island of 348 km^2 with a population (at the time) of 91,000. Just four days after the invasion, the allied Grenadian and Cuban forces surrendered. In the wake of the invasion, many have questioned why the US would deploy such an impressive armada to invade a country that's smaller than even Washington DC?

Among the many reasons offered at the time to justify the invasion, one major argument pointed to the fact that Grenada had strong links to Cuba, the USSR, North Vietnam, and North Korea.[1] It had, in fact, joined the USSR's sphere of influence in 1979 when a coup brought Maurice Bishop to power and the country swiftly transitioned to a communist regime.

Despite four years of criticism of US policies and its hostile presence in the Caribbean (an area considered to be America's "backyard"), tensions were kept under control until 19 October 1983, when Bishop, who was judged too moderate by his party's hardliners, was deposed in a coup and executed alongside some of his ministers and civilians. At this moment, roughly 600 American students were studying at the St George's University School of Medicine. The coup sparked fear among the US government as it evoked the specter of the Iranian hostage fiasco of 1979, just a few years earlier. The invasion also occurred during renewed tensions between the US and the USSR and sparked much international criticism, even from close US allies such as Canada and Britain.

Existing Explanations of the US Invasion

The invasion of Grenada has been studied from various perspectives. The most prominent works are composed of accounts from eyewitnesses and former advisers, historians, and political scientists.[2] Many studies have concentrated on military aspects of the conflict, such as the various roles of the diverse military bodies present to the invasion, the planning and execution of strategies, and the conduct of operations.[3] Others have concentrated their analyses on the legal aspects of the conflict: through the lens of the US Constitution and of the 1973 War Powers Resolution[4]; others have approached it from the perspective of international law, questioning whether the invasion was legal based on the clauses of the OECS Treaty.[5]

Among the most prominent arguments made to explain the US invasion of Grenada are appeals to the US tradition of "gunboat diplomacy" in the Caribbean – in place since the Monroe Doctrine and refreshed in the context of the Cold War. With countries such as Cuba and Nicaragua in the communist sphere, Washington was growing nervous at the idea of the Caribbean becoming a "communist lake."[6] Another notable argument (one used especially by the US administration) was the need to protect the 600 US students threatened by the coup. However, many have questioned whether US nationals were really under threat in this situation.[7] Rather, some have accused that Reagan wanted a regime change in Grenada since his election and was only waiting for a workable excuse.[8] Other arguments advanced include the need to respond to a formal request for assistance made by Caribbean countries, and the need to atone for the humiliation of the Iran hostage crisis of 1979.[9]

In the field of IR theory, the invasion of Grenada has inspired considerably little study, with the notable exception of Cynthia Weber, who analyzes the invasion through a feminist lens as an attempt by the US to re-masculinize itself and assert its (male) domination over a feminized region (the Caribbean).[10] According to Weber, the US had been castrated by their failure in Vietnam, and an invasion of Grenada would be a good opportunity to make America appear virile in the eyes of the world.

The idea of a fight over and for virility also appears in the recent work of Keren Yarhi-Milo, who argues that the invasion occurred because Reagan was what she calls a high self-monitoring hawk or "reputation crusader": the type of individual willing to fight if a struggle involves reputation, even if no material interest is at stake.[11] To prove her "hawk" claim and that the invasion was, therefore, decided on due to a fear of appearing emasculated and weak, she undertakes a computerized textual analysis of US public speeches. Finally, Julian Go's recent research[12] highlighted the need to invade Grenada to respond to US decline and to perform US great power status.

While this study offers insightful findings, it nonetheless poses the important methodological problem of whether the content of a public speech indeed asserts what the speaker believes, or if it is used rather to placate an audience. The remainder of this chapter and the next will attempt, in part, to evaluate these claims by presenting the findings of a mixed method analysis of declassified US documents pertaining to the invasion.

A Priori Evidence Supporting Each Framework

As undertaken in previous chapters, this section aims to provide a priori evidence taken from the existing literature to explain how the three tested theoretical frameworks – realism, the diversionary theory of war, and status-signalling and nostalgic virility – could each be relevant to the US decision to invade Grenada.

Status-Signalling and Nostalgic Virility

This framework is organized around two separate yet often interconnected components: (1) status-signalling and (2) nostalgic virility.

Status-signalling. The idea of going to war because of and for status is probably the framework that has the most a priori evidence to support it. Prior to the invasion, the US had recently suffered two humiliating defeats on the international stage: in Vietnam and in Iran. Despite US withdrawal from Vietnam in 1968, the problems of the war still strongly influenced US politics. Each deployment of US troops became bound up with the fear of a new situation from which it might be impossible to retreat or to win. Adding to these troubles, Carter's presidency ended in complete humiliation in 1979 when he ordered a rescue mission to save US hostages in Iran. Not only did the US lose a key ally and see its position weakened in the Middle East, but it also saw its military humiliated after the failure of the operation that ended in complete chaos. The failure illustrated key weaknesses both in US forces' equipment and in the command and planning chains.[13]

As if these events were not catastrophic enough for US prestige, 1979 was also marked by the successful communist coup in Grenada, bringing Bishop into power, and by the Soviet invasion of Afghanistan. In this context, it was no surprise to see the 1980 presidential campaign focus heavily on the idea of bringing a stop to US decline: "It was candidate Reagan who implied that he would do so [restore US status and credibility as a great power], and do so militarily."[14] Great power status would be found in the maintenance of the world system and international law. Grenada presented a direct challenge to international law and the US was asked to assist weaker Caribbean countries in their attempt to restore order.[15] Given its moral dimension, the responsibility to protect and assist weaker actors also offers us a direct link to arguments based on nostalgic virility.

Nostalgic virility. Building from the evident and latent feelings of humiliation,[16] nostalgic virility can be evidenced by attempts at punishing morally or behaviourally deviant actors. An a priori argument to support the nostalgic virility hypothesis comes from the supposedly barbarous manner in which New Jewel Movement[17] leaders behaved during and after their coup: Bishop and his close advisors were summarily executed, as were civilians, and a twenty-four-hour curfew was established.[18] The US administration was scandalized and, in virile fashion, forced to act against such behaviours. It has been argued that the brutal manner of the coup made Reagan and his advisors decide not to trust Grenadian and Cuban assurances that US civilians were safe. "For better or worse, the Reagan administration decided to ignore the RMC's [Revolutionary Military Council] assurances, believing that the RMC could not be trusted to keep its word."[19]

Realist-Strategic Framework

Realist-strategic frameworks are perhaps the most obvious a priori explanations as to US motives for invading Grenada. Although tiny, Grenada is of huge strategic importance: it is very close to Venezuela, one of the world's largest oil exporters, and two-thirds of US imports transit through the Caribbean. With Cuba already a hostile country, also having a hostile Grenada could prove a threat to US shipping. Even if it lacked the military capabilities to be a serious threat to US military power, it still had much more military capability than its size would require, as the arms stock captured during the invasion later revealed.[20] As well, the construction of a larger airport in Point Salines with the capacity to accommodate Soviet fighter jets and cargo planes increased US suspicions. Ultimately, the US feared that the new airport would be used by Cuba to support its operations in Angola and Latin America.[21]

Other arguments tested are those linked to the context of the Cold War. Much a priori evidence can be found to support the domino theory and the idea of "rolling back" the USSR. With Guatemala, El Salvador, and Honduras already under threat, allowing Grenada to become a communist stronghold would have been seen as a dangerous mistake.[22] In the eventuality of other countries falling to the Soviet sphere of influence, the Caribbean Basin Initiative (CBI)[23] could potentially be at risk.[24]

DTW Framework

Elected for a first term in 1980, Reagan had to face the traditional midterm elections in 1982, during which seats in both Houses of Congress were fought for. Republicans managed to keep their majority in the Senate (a first since 1928) but the Democrats cemented theirs in the House of Representatives by gaining twenty-seven seats. As the US entered the 1982 economic recession, these results indicated that domestic opinion was at best skeptical to Reagan's economic policies. In both his diary and autobiography, Reagan noted the difficulties that the economic recession created for him and his policies, regretting that even his advisors wanted a political U-turn.[25]

With the 1984 presidential election on the horizon, a plan to save "threatened" US citizens in Grenada could have been an easy, quick, and cheap way to improve domestic popularity just a few months before the official start of the campaign. It might also have helped Reagan in his relationship with the Democratic-led House of Representatives, as in US politics both the president and congress are bound to collaborate and compromise in order to pass bills. A victory would strengthen his position by making his leadership stronger. Furthermore, Reagan had been elected on the explicit claim that he would stop the US decline: "A bold move was required for the President's credibility."[26] Accordingly, a number of a priori arguments can be made to support the diversionary theory of war view for the invasion of Granada.

Having briefly outlined some of the a priori arguments and perspectives from the existing literature, the remainder of this chapter will outline the findings in support of the realist-strategic and diversionary theory of war frameworks, respectively, before turning to the evidence for the status-signalling and nostalgic virility frameworks in chapter 8.

Evaluating Realist-Strategic Framework: A Relatively Strong Presence, but Detached from Grenada

Economic Interests? Yes, but Marginalized and Tied to Regional Geopolitical Concerns – Not to Grenada

Analysis reveals that the situation was described more in strategic terms (20.6 per cent) than in economic ones (5.5 per cent). These initial results lead us to assume that economic arguments made in the literature (e.g., surrounding sea lanes, exploitation of oil, etc.) did not weigh heavily in the decision to invade Grenada. Explicit economic concerns and arguments were difficult to find in the dataset; although present in 1.5 per cent of cases (six documents in the entire dataset), if we divide the category of "economic concerns" into the codes that compose it, the results appear even less significant. Namely, the codes potential gain/loss of natural resources, potential trade gain/loss, and potential commercial gain/loss appear in 0.5 per cent, 1 per cent, and 0.5 per cent of cases, respectively. The idea of protecting economic interests remains stable in *Phases 1* and *2* (1.3 per cent) and only increases at *Phase 3* (2.6 per cent), after the invasion.

Qualitative analysis of the documents where these codes were found reveals a few key pieces of information. First, economic arguments were not found in important documents, only in-service notes and memos, and never in documents to or from high-ranking officials. Second, Grenada was not explicitly mentioned. Instead, these documents make reference only to the Caribbean region generally.[27] If a few countries were explicitly cited as being of importance, the list did not include Grenada, but rather those such as Cuba, Mexico, Nicaragua, Panama, and Venezuela. Third, the qualitative analysis of the documents where codes related to economic concerns appear shows that the economic interests were not self-sufficient argumentative logics. In other words, they did not appear on their own, but in close connection to broader geopolitical and strategic discussions: "The major shipping lanes crossing the region make it one of our major lifelines to the outside world, and, as a result, an area of crucial importance to the continued prosperity and security of the United States. The defence of the Caribbean, however, is complicated by hostile forces in Cuba and Nicaragua."[28]

Geopolitical Motives: Containment and Roll-Back – Yes; Planned Invasion – No

The election of Ronald Reagan as president of the US in 1980 marked a new phase in the Cold War, a phase of renewed tensions as the newly elected president was determined to put an end to global communist expansion. Based on this fact, and on the relevant literature, I expected to find a strong Cold War-orientated strategic narrative in the dataset. The results show that the situation was described in strategic terms in 20.6 per cent of cases. Furthermore, a military-technical vocabulary was found in 25.1 per cent of cases and evidence supporting an explicit geopolitical framework in 18.3 per cent. To measure the relative importance of the geopolitical component of the strategic-realist framework, I tested several codes. Given the Cold War context, I first coded for evidence of the domino and roll-back theories, followed by more traditional geopolitical ideas (e.g., gaining or losing a bridgehead, a position used to support actions in another area, or a buffer zone).

Cold War Geopolitics and the Grenada Invasion: Two Different Sets of Villains to Contain and Roll-Back

So, how was the geopolitical narrative organized and what was it composed of? Which narratives were revealed as dominant? The analysis shows the relative importance of a Cold War-oriented strategic narrative built around two typical arguments: domino-effect theory (11.6 per cent) and rolling back the adversary (8.5 per cent). The domino-effect narrative and argumentation was supported in some of its strategic implications by a secondary, yet low-frequency, narrative. This secondary narrative was built around the ideas of a gain/loss of a strategic location (2.5 per cent), a gain/loss of supporting locations for action in other areas (0.8 per cent), a gain/loss of buffer zones (0.3 per cent), and a gain/loss of bridgeheads (0.8 per cent). Together, these minor yet important strategic considerations helped shape the domino effect theory as a key argumentative logic in discussions over Grenada.

How did this narrative develop? Through qualitative analysis, it appeared that the Caribbean was described as a region open to communist subversion and influence mainly because most of its states were simply too weak to check Cuba and the USSR: "The emergence of a Cuba/Grenada/Suriname axis is of major geopolitical importance, particularly in the relative absence of countervailing forces."[29] In addition, it was thought that both Cuba and the USSR were trying to increase their capabilities in the region by helping Grenada expand its existing airport: "Moscow may also have helped finance some seven hundred Cubans on the island, the majority of whom were working on the Point Salines airport."[30] Despite claims that it would only be used to increase tourism, it was feared that the airport would be used by Cuban aircraft to resupply on route to Angola, as well as by Soviet fighters.[31] Nonetheless, the situation was not judged to be of concern until *Phase 2*.[32]

During *Phase 1*, intelligence reports issued by the CIA rejected the possibility that the Soviets would try to implement nuclear missiles in Grenada, as they had attempted to do in 1961 in Cuba: "We do not believe Moscow envisions conducting substantial naval operations to cut such sea lines of communication in wartime. Nor do we think Moscow sees either country [Grenada or Suriname] as an attractive base for Soviet ballistic missile submarines or land-based ballistic or cruise missiles."[33]

The level of (perceived) threat was raised after news of the coup became known. The new regime was more radical and friendly toward Moscow, and senior US official Robert "Bud" MacFarlane wrote to Reagan that this time there was a possibility of seeing Soviet SSBNs deployed in Grenada: "The port of St George's is capable of accommodating Soviet nuclear ballistic submarines and their tenders – another possible Soviet response to INF [Intermediate-range Nuclear Forces]."[34]

The domino effect narrative remained peripheral and relatively unimportant until *Phase 2*, when discourses referring to the idea of rolling back Soviet influence also began to appear with more frequency. The more proactive idea of rolling back communism appeared in *Phase 2* of the conflict (13.4 per cent of documents), linked with a call for help from Caribbean states to the US. "In order to stop the carnage and spread of communism, I recommend that you positively support the call for help by our friends and neighbours," wrote the US Ambassador in Bridgetown for the attention of Reagan on 22 October 1983.[35] To better understand how these narratives played out and shaped perceptions and decision-making, it is helpful to consider which actors were pegged as the "villains" of the conflict.

For example, the domino effect narrative was animated by reference to two sets of "villainous" actors requiring containment. The first group was the communist bloc, perhaps obvious given the historical context (and given that the Grenada invasion was identified as part of the global fight against communism in some 29.1 per cent of cases). What did come as a surprise was that, despite the clear framing of the Grenada invasion as part of the East-West struggle, Cuba and the USSR were not often identified directly as villains (only in 4.3 per cent of documents overall). However, their direct identification as villain did increase over time, from 2.7 per cent in *Phase 1* to 7.8 per cent in *Phase 3*. The second group of "villains" is perhaps less obvious. *Phase 2* marked the moment when domino effect and rolling back the adversary narratives began to be associated not only with communism or Cuban and Soviet influences, but also with the chaos in Grenada after the coup. The new "villain" emerged as the chaotic regime in place in Grenada, thus expanding the scope of the realist-strategic framework out of the singular context of the Cold War. Not only was the region threatened by the spread of communism – something that needed to be contained and then rolled back – but now also by the internal political chaos that also had to be contained. This new framing was used in combination with the Cold War narrative – as seen in the diplomatic cable quoted above – as well as independently, as seen in the following diplomatic cable sent by the Secretary of State to the US delegation at the UN: "The dangerous vacuum of authority now existing in Grenada, which shows no sign of correction, has seen deemed … to be a threat to the continued peace and security of the Eastern Caribbean."[36]

Overall, then, we find that strategic narratives surrounding the Cold War and communism, as well as more localized concerns over the political regime in Grenada, helped shape perceptions and drive a narrative pegging a proactive response to threats and chaos in the region as a necessity. We may tentatively conclude that realist-strategic ideas thus played a role in shaping the response and overall decision to invade. However, important questions still remain. Namely, given accusations in the literature and the press (at the time) that Reagan had planned to invade Grenada since the beginning of his presidency, what does the evidence suggest? When did the idea of invading Grenada become a serious consideration and begin to concretize?

A Planned Invasion since the Beginning? Not Quite Sure

Confronted with this "chaotic" situation and the threat of communist expansion, how did the US react? There were two opposing scenarios. The first was defensive and consisted of containing the threat; the second was offensive and meant rolling back the opponent. The literature is divided on this. Some scholars argue that the Grenada invasion was planned long before the coup, even since Reagan was elected president.[37] According to scholars defending the hypothesis, the strongest proof is that: "the Pentagon with its mandate to rattle communist nerves in the region began practicing military manoeuvres in the area. A US navy exercise dubbed 'Ocean Venture 81' included a scenario involving the rescue of American hostages on the fictional islands of Amber and Amberdines."[38] The implicit reference to Grenada and Grenadines is evident, and the military exercise was one of the largest ever seen, as 120,000 troops, 250 warships and 1,000 aircraft were deployed.[39] This claim was also supported by the *Los Angeles Times* in its 26 October 1983 article about the Grenada invasion: "The Reagan administration from its first days has made no secret of its concerns about Grenada as a new and threatening Cuban and Soviet outpost."[40]

To assess whether or not this claim is correct, I coded for direct and indirect evidence indicating that the invasion was planned before the coup. Direct evidence was gathered by measuring the frequency with which each possible outcome to the situation in Grenada was discussed: namely, either negotiating with Bishop, following a hardline, or refusing to negotiate. I then restricted the calculus only to internal documents from *Phase 1*, because if a military intervention was planned before the October coup, it is at this stage that we should expect to find most evidence. By the same logic, it should surely be found in internal documents of the US administration. For explicit evidence of a planned invasion, I looked at possible favoured outcomes to the situation in Grenada by members of the administration. The results are shared equally between the idea of negotiating with Bishop (9.1 per cent) and the idea of following a hard line (9.1 per cent), while the third option of refusing to conduct negotiations was present but lower frequency (4.5 per cent). I proceeded to qualitative analysis of the documents to investigate the hardline arguments, although the analysis revealed that no invasion was being planned and that the hardline option did not include military invasion of the island.

Right from 1981, the administration was somewhat confused over which policy to follow. Reagan's assistant, Richard Allen, complained in a memo to Roger Fontaine, a member of the National Security Council (NSC) and Director of Inter-American Affairs Department, that "The paper is not clear as to what the ultimate objective is. Are we content to contain Grenada ad infinitum or bring down the regime?" He continued by suggesting that the best option would be to avoid any direct action so as not to "offer a target, let others publicly denounce, don't give Bishop a chance to be a Martyr on the cheap."[41] If regime change was envisaged, it was preferable that it came from the inside and not from military invasion. This strategy did not seem to change across time, as evidenced in 1983 by a memo addressed to Reagan by William Clark.[42]

As the explicit evidence was not decisive, and as it seemed to go against the idea of a planned invasion, I turned to indirect and implicit evidence using two codes. The first, deliberate conflict initiation, was applied to any reference to the invasion having been planned before the coup. The second, dragged into conflict, was applied to any reference to the US having no choice and being forced to intervene. Overall, dragged into conflict appeared in 14.1 per cent of cases and deliberate conflict initiation in just 1.3 per cent. The idea of being dragged into conflict is found in 15.1 per cent of internal documents, while that of deliberate conflict initiation never appears internally. Similarly, in *Phase 1* the idea of preparing the invasion was never identified.

In order to double-check the results, I qualitatively analyzed the few documents in which the idea of a planned invasion was identified. All hits came from newspapers and were accusations from journalists. However, the argument of a planned invasion cannot be definitively ruled out. Very few documents regarding the exercise "Ocean Venture 81" are available as all documents about its planning and objectives are still classified for national security reasons. In this context, it could be possible that the exercise was planned to prepare for invasion, but until further documents are made available for research it is a claim that cannot be asserted solely on current data.

The solution that was planned, at least until the coup, was to strengthen other countries in the region to ensure they were best able to resist communist subversion. US Under Secretary of State, Lawrence Eagleburger, recommended the implementation of a containment policy "which does not attempt to address the fundamental problem of Cuban and Soviet involvement in the Eastern Caribbean, but it may harden the targets of Cuban subversion."[43] In a document dated 1981 (no specific day), the US military seemed to take a similar line as it did not recommend direct military action against Grenada but instead hoped to "encourage Grenada to return to democratic rule, the constitutional process and a Western orientation, or at least, genuine non-alignment."[44] In the same document, a careful and discreet approach was also recommended. The evident power disparity between the US and Grenada was judged unfavourable to the former. Even criticism needed to be judged carefully: "More active US public diplomacy may further isolate the PRG but it would have to be carefully handled to avoid being counterproductive. Too strong US criticism … could play into Bishop's hands by giving the impression of American bullying which would earn him some sympathy in the region."[45]

The Realist-Strategic Framework: A Temporary Conclusion

My analysis has evidenced the presence of a realist-strategic framework built around two main narratives: fighting communism and containing chaos. No real, serious economic interests were detected and the idea that the US administration had planned the invasion of Grenada since 1980 has to be rejected, based on currently available data.

Nevertheless, the answer to the question *Was the realist-strategic framework the first wheel that put the invasion process into motion*? is "probably not." I believe it is highly likely that the framework was largely a post-fact narrative used by the US administration to justify its decision to invade to the American public (which will be discussed below).

Evaluating DTW: Implicit and Explicit Articulations of the Framework

A Weak Framework or Puzzling Results?

If the decision to invade Grenada was made to increase domestic popularity, we should expect to find ideas and arguments surrounding the risk of losing power, losing elections (legislative, presidential, etc.), or losing domestic support (e.g., rising hostility to the administration, risk of losing support of electors). Throughout the coding process, very few relevant codes were applied as these aspects were so seldom mentioned in the dataset. It was unsurprising, therefore, to find that evidence for the DTW framework appeared in only 1.8 per cent of documents. Of all the codes relative to the framework in the coding dictionary, only the following were used: concerns about losing presidential elections (0.8 per cent), concerns about losing parliamentary/legislative elections (0.5 per cent), concerns about poll ratings (0.3 per cent), concerns about parliamentary/legislative hostility (0.3 per cent), and rallying domestic opinion around the flag (0.8 per cent).

Despite the low frequencies, these arguments could still be relevant if there was an increase of frequency from *Phase 1* to *Phase 2*. If there were an increase, it might suggest that there was at least concern about popularity or US domestic opinion when making the decision to go to war; however, it was not the case, as the frequency decreased to just 1.3 per cent in *Phase 3*. Finally, the framework appeared mostly in public documents (5.5 per cent), with a small exception of 0.4 per cent in internal documents. This tells us two things. First, the US administration did not use the framework to any meaningful degree, which indicates that their political fate was not a concern or that they did not feel politically threatened. Second, the framework was used by the administration either publicly for political purposes or by the public itself. It appears to be used only by advisors (2.2 per cent) and public (7.1 per cent) actors. As Reagan's political advisors used the framework (even if infrequently), I proceeded to a qualitative reading of the documents to double-check if the low frequency may be misleading. The fact that the "public" had the highest frequency could imply that it used charges of diversionary intentions as a way to criticize the way Reagan and his administration were managing the situation in Grenada, or simply to criticize his policies more broadly.

Political Use of DTW: Criticizing Reagan for Diverting US opinion; Views from the Press and Congress

To know which elements of the DTW framework were used by the US public, I first investigated codes related to the variable public. Results show that the US public believed that Reagan's administration had concerns about presidential elections in 2.9 per cent of documents, about

Table 7.1 The use of DTW-related codes by US public across time

Codes	*Overall*	*Phase 1*	*Phase 2*	*Phase 3*
Concerns about presidential election	2.9	11.1	2.4	0
Concerns about parliamentary elections	2.9	11.1	2.4	0
Concerns about poll rating	1.4	11.1	0	0
Rallying domestic opinion around the flag	4.3	0	0	5
Making domestic opinion forget about something	2.9	2.9	0	0
Own administration as villain	17.1	11.1	19.5	15

parliamentary/legislative elections in 2.9 per cent of documents, that poll ratings were an issue for them in 1.4 per cent of documents, and that Reagan was trying to rally domestic opinion around the flag and make domestic opinion forget about something in 4.3 per cent and 2.9 per cent of documents, respectively.

In addition, the US public clearly identified Reagan and his administration as villains in 17.1 per cent of cases. By comparison, the public identified Grenadian leaders as villains in only 10 per cent of cases. I was curious to see if there was an increase based on the evolution of the situation in Grenada and so computed the DTW codes with the variable *actors*, successively associated to the variables *Phase 1*, *Phase 2*, and *Phase 3*. The results show there was an evolution, but not in the way I had envisaged, as the frequencies completely eroded from *Phase 2* onward (see table 7.1).

Based on these results, we can argue that the narrative used to criticize the US administration would probably have been built around the idea that Reagan was planning to invade Grenada in order to win a quick and easy war to bolster public support. To better understand which segments of the category public were driving this narrative, I analyzed the three components of the category separately: congress, domestic opinion, and the press. The DTW framework was evidenced by the press in 6.7 per cent of cases and, added to it, the Reagan administration was framed as a villain by the press in some 16.7 per cent, indicating considerable levels of critical narratives circulating in the public sphere. Similar DTW-related ideas were also used by congress in 10 per cent of documents, with a considerable 20 per cent of documents also portraying the Reagan administration as villain. However, in the few documents sent by US citizens (considered here as domestic opinion), no DTW-relevant evidence appeared.

So, what can we take away from these findings? For starters, much of the press never really bought into the Cold War narrative of Reagan's administration, according to which Grenada was becoming a serious threat because of Cuban and Soviet involvement. On 26 October 1983, two days after the US forces landed on Grenada, an unnamed columnist from the *Hartford Courant* expressed doubts about US strategic motives: "If what was happening in Grenada became so untenable, why did the government of Venezuela, the largest nation in the region, not participate in or even lead the rescue mission? Why did the government of British Prime Minister Thatcher advise against invading the former British colony?"[46] Questioning the Reagan administration's real intentions towards Grenada was nothing new. Since rising to power, he had received letters from concerned politicians asking him not to invade.[47]

The point raised valid questions surrounding Reagan's argument that the airport would have a military use, particularly because Cuba and the USSR were not the only actors helping Grenada finance the construction: they in fact included the UK, Canada, and even private US investors. Given Thatcher's well-known attitude towards communism and the Soviet Union, it would be extremely awkward if she had agreed to finance an airport that might present the risk of being used militarily by the Soviets.[48] To all these letters, the White House response was always the same: "You may be assured that the United States government is not planning to invade or destabilize Grenada."[49]

The other widely used argument was to question Grenada's actual military capabilities: if Grenada was so heavily armed, where was the proof? An article in the *New York Times* on 26 October 1983 (unknown author) illustrates the point: "If President Reagan deserves the benefit of any doubt in Grenada, it is the possibility that Cuba and the Soviet Union had indeed moved, with only modest investments of men and weaponry, to establish a puppet regime that would give them bases for other operations, but if that were clear, why was it not proved, or even asserted?"[50] Not only was the Cold War strategic narrative questioned, but so was the reality of the threat that US students were supposedly facing. Protection of US citizens in Grenada was the other major justification given for invading Grenada (18.1 per cent of documents; how this narrative was used and constructed will be developed in chapter 8), although the authenticity of this narrative was also frequently questioned, for example in the public commentary of the invasion by Representative Peter Kostmayer: "If Americans are in danger, the United States has an obligation to do all it can. The question is were they in danger or was this used as a pretext for the invasion?"[51]

The press was, of course, speculating about this, too. While the Reagan administration faced problematics from the negative testimonies of some rescued students – such as Anthony Turner, who claimed: "I never felt in any danger until the invasion actually happened"[52] – such remarks were rare. Of all comments by students available in the archives, only the one statement by Turner is negative; all the others say that they did in fact feel as though they were under imminent threat. It seems that those who questioned Reagan's narratives, then, were essentially the press and congress. According to various polls, US domestic opinion was mostly convinced that students were in actual danger (51 per cent for, 31 per cent against, 18 per cent with no opinion).[53]

As the Reagan administration's narratives were not believed by everyone, there was speculation about their genuine motives. The press thought the real reason was "to get a quick and easy victory"[54] because "hostility toward the White House is approaching critical mass, that point at which the chain reaction becomes self-sustaining."[55] Meanwhile, another columnist explicitly made the claim that Reagan ordered the invasion to gain public support and secure his re-election: "As Thatcher showed with the Falklands/Malvinas dispute of last year, there is nothing that rescues a politician in trouble like a dandy little war against someone who's considerably weaker."[56] Nonetheless, these speculations help animate and contextualize the overall story, yet offer little in the way of proving the motives that led to the decision to invade Grenada.

In summary, the findings show that the opposition to Reagan never believed in his justifications. To them, Reagan invaded Grenada for political purposes only: to bolster his position for the next elections. As such, Reagan was often portrayed as the villain of the story because he created, "out of nowhere," an international crisis to serve his personal political objectives. However, these public narratives and speculations find little to no evidence or support in the declassified documents analyzed for this study. As such, we cannot credibly support a DTW-based interpretation of the decision to invade Grenada. Nonetheless, before concluding it is worth briefly outlining an interesting hybrid between the realist-strategic Cold War framing and a post-fact justification which took place after the invasion in a clear attempt to gain public support.

The Surprising Use of the Cold War Narrative: Gaining Post-Invasion Public Support by Increasing the Perception of Military Threat Posed by Grenada

In my analysis of the strategic-material framework I hypothesized that the Cold War narrative was used as a post-invasion justification, probably to counter-balance criticism about the invasion and attempt to gain public support by proving that there had been a real threat to US students and to the region more broadly. This claim was suggested by the frequency distribution based on the phase variable. In *Phase 1*, the Cold War geopolitical narrative was found mostly in the administration's internal documents (13.1 per cent compared to 10.5 per cent for public documents). Yet at *Phase 2*, the frequency curves reversed and the framework was more present in public documents (25.7 per cent compared to 13.2 per cent). It is only at *Phase 3* that the curves reach more or less the same level again: 26.3 per cent in internal documents and 25.7 per cent in public documents. What interests us here is the frequency at *Phase 3*.

The Cold War narrative was still being used frequently during and after the invasion, especially in public documents (25.7 per cent); the administration used the idea of a global struggle against communism in 31.3 per cent of cases and the Cold War geopolitical narrative in 18.8 per cent. This indicates that the administration believed it was in their interest to continue using these ideas in its public discourses as the public were quite receptive, using ideas related to the Cold War geopolitical narrative in 30 per cent of public documents and that of the global struggle against communism in 45 per cent (i.e., the combined use of the variables public and *Phase 3*).

To determine whether the Cold War geopolitical narrative was used as a post-fact political support mechanism, I tracked how the public and the administration used these narratives. In the case of the public, there was no real evolution: the ideas were used to criticize and question the motives of Reagan and his administration (as previously detailed). Qualitative reading of the documents where the administration used these narratives is much more interesting and reveals something new. It shows that the administration planned to display all the "captured" documents and military equipment in Grenada and to organize mass media coverage of the display. In a weekly report to Robert McFarlane, Diane Dornan explained that this decision had Reagan's complete support: "The President's interests in displaying captured Cuban material … suggested that we bring all the material captured in Grenada to Washington and display it with press coverage … He was very taken with the idea."[57] A draft of a public action plan at the end of the military intervention in Grenada (date unknown) reveals that, to fight "the perception of US gunboat diplomacy – lack of threat to American citizens," it was recommended to "utilize captured documents and other evidence as soon as possible."[58] We do not know if this objective and formulation was kept in the final version of the plan. What we do know is that the total number of seized weapons and equipment displayed to the public was impressive and illustrated Grenada's strong ties as it "included … small arms, machine guns, ammunition, armored personal carriers, and other military equipment of Soviet, Cuban, and North Korean origins."[59]

Before the seized documents and equipment were displayed, they were heavily annotated by military staff and the administration. Every comment emphasized how disproportionate the equipment was for such a tiny island, and that this could only indicate one thing: they were not meant to remain in Grenada but to be used in other countries. The 31 October briefing for the Freedom Federation and ethnic leaders highlighted that "85,000 weapons, one for every man, woman, child on the island [were found]. In addition, 12,000 army uniforms were discovered, even though the Grenadian militia numbered only 2,000 men."[60] US forces that invaded Grenada declared surprise at how heavily armed Grenadian and Cuban troops were. The 9 November 1983 memo for the Senate Committee illustrates this well: "To put this quantity of ammunition and weapons in perspective, the commanding general of the 82nd Airborne repeated that he brought only 275,000 rounds of small arms ammunition to support 2 brigades (or about 4,500 soldiers) compared with 5,500,000 rounds found … the number of mortars captured exceeded the number authorized for the entire 82nd Airborne division."[61] What can we take away from these findings, then? On the whole, we can confirm that the DTW framework does not find support in the data as a strong influence on the decision to invade Grenada, but that a post-fact hybrid of the framework, using the Cold War strategic narrative, was used by the Reagan Administration to gain public support only after the end of the military operations. This served to convince the US public that the administration made a good decision in invading the island, despite early criticisms and doubts expressed. And in this sense, this realist-strategic narrative may have served a DTW-related purpose, something worth exploring in more depth in future research. Nonetheless, given that this public support display occurred after the fact, we cannot posit it as an influence on the actual decision to invade Grenada, based on currently available data.

Conclusions on DTW

Was the DTW framework important in the case of the 1983 US invasion of Grenada? While we cannot make any definitive conclusions about the theory – and further research is certainly necessary – the results of the analysis suggest that this framework had a very minor presence but may have played a small role of some sort in the overall decision-making process. The low frequency of explicit evidence for the framework as part of the administration's decision-making would lead us to conclude that it was unimportant. However, this is in fact balanced by the presence of a much stronger indirect framework among the public that suggests domestic opinion may have played a larger role in influencing the decision to invade Grenada and the Reagan administration's responses and justifications thereto: namely, that Reagan and his administration were not willing to lose support during and after the intervention and tried to convince the public that they were right, that Grenada posed a real threat that had to eliminated. All these hypotheses remain speculative so long as new classified documents await release. Overall, however, we may conclude that the realist-strategic framework played an important, if not overly dominant role in the decision to invade Grenada, while the DTW framework may have served more as a PR strategy and dynamic in the public discourse than as an influence on the decision to invade.

8

The US Invasion of Grenada, 1983 (II)

Testing the Status and Nostalgic Virility Framework

This chapter continues the investigation of the motives leading to the Reagan administration's decision to invade Grenada in 1983. A priori evidence for the status and nostalgic virility frameworks was presented in the last chapter. Here, the empirical evidence for these frameworks will be evaluated and placed in conversation with the evidence for realist-strategic and DTW frameworks outlined in chapter 7. We will begin with an evaluation of evidence for the status framework, looking at questions of great power status and status signalling, followed by an evaluation of evidence for the nostalgic virility framework. By the end of this chapter, we will have a more holistic understanding of the perspectives and motivations leading American decision-makers to engage in war over Grenada.

Evaluating Evidence for Status: A Caribbean Melos in the Making?

In 416 BC, during the Peloponnesian war, Athens besieged the tiny island of Melos located 110 km east of mainland Greece. Athens demanded Melos surrender and join her empire or face complete destruction. Melos refused and was, indeed, annihilated. Thucydides, in his *History of the Peloponnesian War*, explains the reasons behind the argument through a dialogue between the Athenians and the Melians.[1] The latter argued that they were too weak to cause any threat to Athens, and thus it was not necessary to destroy them. The Athenians replied that they were aware of this but could not let Melos be independent for symbolic reasons and the negative signals it would send. Athens' empire was based on its control of the seas: if Melos was not conquered, it would signal that Athens was weak and increase challenges against it. Melos must be destroyed, not for what it was or threatened in material terms, but for the symbolic risk it represented. This section will investigate whether or not there is evidence to suggest that the invasion of Grenada was a modern-day repeat of this ancient status-signalling logic – a Caribbean Melos, of sorts.

Broadly speaking, mentions of status and its associated symbolism, both implicit and explicit, appeared in some 50.1 per cent of documents within the dataset. Nevertheless, specifically status-related arguments surrounding the conflict and invasion (i.e., those clearly engaging the status-signalling theoretical framework laid out in chapter 2) appeared in 26.6 per cent of cases. Based on these initial frequencies, we see that questions of status played an important role within the ecosystem of ideas influencing and circulating between decision-makers during the conflict. Every category of actor used status-related arguments at more or less similar frequencies: Reagan (34.1 per cent), secretaries of state (32.3 per cent), Reagan's advisors (23.9 per cent), and the public (32.9 per cent). Reagan appeared to be the most status-prone individual, which is no surprise as he was elected on the explicit promise to stop a perceived US decline and re-establish America as the world leader. As with the UK case, the second most status-concerned group of individuals was the public, a finding which could support an interpretation of the invasion as a variation of the DTW, one that would be a modern version of the Roman practice of evergetism. If this is the case, we should find explicit and tangible proof that the public had a clear idea about the US status and its role(s) on the international stage. Finally, if the status-signalling framework is relevant to this conflict, then we should expect evidence showing that status is both gained and kept through recognition by relevant others (e.g., other great powers).

Interestingly, when calculating the status-related codes they showed a higher frequency in external documents (35.7 per cent) and public documents (33 per cent) than in internal documents (24.4 per cent). Following this frequency distribution, we may hypothesize that the frequency obtained within external documents indicates that the US tried to reassure other countries about its willingness to fulfil great power duties, and that its decline had stopped then and there. The frequency obtained by public documents would reinforce the "bread and blood" argument of evergetism. If true, we should find the following: (1) claims of performing great power duties in Grenada and of sending appropriate status signals, and (2) selling the war to the public by using status-related narratives. These hypotheses will be investigated throughout the first part of this chapter.

First, however, we must ground the analysis of status-related discourses by investigating how US actors understood their own role as a great power. Considering the initial frequencies above, there seems to have been a general consensus among actors that status was an issue or was at stake in the conflict. Surprisingly, however, evidence of self-attributions of a specific status was found in only 5.8 per cent of documents. Nonetheless, a close qualitative reading of these few documents can help us answer the question: *How do US actors frame or understand US status, and do they define roles for the US surrounding its status as a great power?*

The US Understanding of Great Power Status: No Status without Responsibilities

Strong Military and Management of the International System: The Heart of the US's Understanding of Great Power Status

After coding every explicit reference to what it means to have great power status and on what it is based, the results were a bit disappointing, quantitatively speaking, with only a handful of references identified in some 3.8 per cent of documents. In the case of the US, great power status was conceived first and foremost around the possession of strong military capabilities (2 per cent of documents), which is unsurprising given the military importance it acquired after the two World Wars. Having a strong military was judged as the sine qua non condition for the US to achieve its objectives: "The second essential precondition of a strengthened and purposeful foreign policy was the rebuilding of the foundation of our military strength ... We have been compelled to make the reconstruction of America's military capabilities another of the highest priorities of this administration."[2] As such, Reagan's administration actively tried to rebuild America's military: "During the past two years, the amount of gross national product going into the military has risen from $131 billion to 182 billion, reflecting the president's belief that if America is to regain its greatness, it must rebuild its military force."[3]

This understanding of great power status's criteria is backed up by a narrative expressing a feeling of military superiority, which was found in the many letters that were sent to Reagan to thank him for his decisions in Grenada and to praise the US military: "Army Rangers and 82nd airborne division acted swiftly, professionally and courageously, they were the greatest sights since leaving US soil."[4] Although used to criticize Reagan's decision to invade Grenada by illustrating the power disparity between the US and Grenada, the feeling of superiority built around military strength was also present in the press, which referred to the US as "the most powerful nation on earth."[5]

Beyond the few minor mentions of what constitutes or leads to great power status, there were also several mentions of the privileges and responsibilities that it entails. Namely, great power status was explicitly linked to the perceived responsibility to manage the international system – solving international crises, protecting and enforcing international law, etc. – in some 15.6 per cent of cases, where it was especially used as an explicit justification for the invasion. The argument for managing international law and order was mainly premised on the idea that the principles of international order (i.e., basic human rights and democracy) were threatened by the events happening in Grenada.[6] Another formulation of this justification and responsibility to manage the world order was given by the US Ambassador in Bridgetown, Lewis Lehrman, to the secretary of state: "To prevent further bloodshed in Grenada and to restore that island to democracy."[7] The responsibility to manage and protect international order was also expressed in a different way, connected directly to the Cold War and irrespective of Grenada itself: "This [fighting against communism] is not a partisan issue. It is a question of us making our moral responsibility to ourselves, our friends, and our posterity. Who among us would wish to bear responsibility for failing to meet our shared obligations?"[8]

The logic behind this argument is that if the world order was under communist threat, then communism had to be tackled everywhere it appeared, including in Grenada. Other generally accepted criteria of great power status, namely having interests in other regions than the homeland, were not explicitly mentioned in the available dataset. There was mention of the US having global interests, but locations were not clearly identified.[9] Interestingly, behind the US's understanding of great power status were two well-established and identified narratives that I will analyze in the following sections: the US's role as leader of the free world/world police, and as protector of democracy.

Responsibility to Accept Self-Associated Roles with Great Power Status: Two Roles for One Narrative

As the discussion of perceived responsibilities above shows, the idea of having specific roles on the international stage was found quite regularly in the dataset (11.3 per cent overall), and it was built around two main narratives. The first was being the leader of the free world/world policeman; the second was being protector of democracy. These narratives were also reinforced by the presence of two others that illustrated, according to the US, that the Grenada conflict was also an expression of two broader struggles: democracy versus dictatorship (11.1 per cent) and rule of law against authoritarianism (8.5 per cent). Both the leader of the free world/world police and the protector of democracy narratives became more frequent after the news of the coup and Bishop's execution: 8.1 per cent of cases in *Phase 1*, rising to 12.8 per cent in *Phase 2*, and 16.9 per cent in *Phase 3*. The increasing frequency was to be expected as the provocation created a context demanding a US response, especially as Caribbean countries explicitly called for US help. The narratives were also used most often by Reagan (27.3 per cent), while all other actors used them at a lower frequency (8.7 per cent for advisors; 11.4 per cent for the public; 12.1 per cent for members of the administration). Finally, they were found more frequently – as expected – in external documents (21.4 per cent) and public documents (14.7 per cent) than in internal documents (9.2 per cent). So how did these narratives play out over time and influence decision-making?

In *Phase 1*, the idea of the military being used as a means to support democracy had already been found, as shown by Reagan's public intervention at a news conference on 26 July 1983: "We support democracy, reform and human freedom ... And, yes, we support a security shield for the region's threatened nations in order to protect these other goals."[10] In *Phases 2* and *3*, the use of the military to fulfil these roles was no longer simply suggested but rather presented as an absolute requirement and even as the appropriate method to follow. A letter addressed by Brigadier General Jose Bloise to Reagan illustrates this: "In 1965, the US sent the Marines into the Dominican Republic to help establish a free democratic government ... It is historical proof that it is important to use the Marines and it does produce beneficial effects to the people of the government in need."[11]

After establishing democracy, the use of the army was seen in *Phase 3* as the only way to guarantee the survival of the newly established government in Grenada.[12] The role of being world policeman/leader of the free world was associated not only with the idea of protecting democracy but also to that of military power, as can be seen in a letter Reagan sent to Thatcher a few days after Bishop's execution: "The members of the Eastern Caribbean States (OCES) have raised similar concerns which are, of course, magnified by their proximity to and limited ability to defend themselves against the threat which Grenada now poses."[13] As the world's policeman, the US had the task of protecting the weak. The idea of protection of the weak was found in 18.1 per cent of cases and will be analyzed in depth below in the section dealing with the virility framework, with which this code has very close connections.

There is one final element about the self-attributed roles of world policeman and protector of democracy that is worth looking at. Emphasis was placed on different aspects of the narrative depending on the audience. For the US public, the narrative emphasized the role of leader of the free world (12.8 per cent) rather than protector of democracy (5.5 per cent). However, when the audience was other heads of state or heads of government, it was the opposite: the narrative centred around being the protector of democracy (17.9 per cent) and not on being the leader of the free world (3.6 per cent). A possible explanation? Being the world policeman/leader of the free world was used to please and reassure the US public, and more specifically Reagan supporters, to whom the idea of being the most powerful country in the world was important. This idea was not used in correspondence to other heads of state and heads of government simply because such a narrative might prove counterproductive to gaining support, especially as US presence in the Caribbean and in Central and Latin America was often associated with imperialism.[14] Instead, the idea of protecting democracy would appear less belligerent.

A Divide over the Best Status Signals to Send: To Invade Grenada or Not?

A key aspect of the status-signalling framework investigated here argues that status is gained through recognition by relevant others, which requires the sending of appropriate status signals. To determine whether such relationality was present in the minds of US decision-makers during the US invasion of Grenada, I coded for any implicit or explicit markers acknowledging it and for any debates over which signals were appropriate to send in order to gain status recognition. Overall, such markers were identified in some 10.8 per cent of documents in the dataset. The frequency increased from *Phase 1* to *Phase 2* (11.4 per cent to 12.1 per cent) but decreased in *Phase 3* (9.1 per cent). Among governmental actors, these ideas were most frequently used in advisors' documents (13 per cent), although the highest frequency of all actors was in fact found among public documents (22.9 per cent). Qualitative analysis reveals a strong divide in the debate over which status-signalling approach to use, namely a partisan-political opposition as to how to respond to the coup in Grenada. This section will investigate this debate to help determine how decision-makers' choices were (or were not) impacted by the question of status-signalling.

Not Invading Grenada to Protect America's Reputation: The View of the Press and of Democrats

The necessity of sending appropriate status signals was acknowledged by the press and in congress. Overall, the press was at best skeptical about the need to invade Grenada, and congress (essentially the Democrats) also raised concerns.[15] It is thus unsurprising that a counter-narrative to the Reagan administrations' pro-invasion rhetoric developed, in tandem with other level-of-threat-based narratives used to criticize Reagan (investigated in chapter 7), based on the idea that any "aggressive" US policy would be damaging to its image, reputation, and claimed roles (i.e., protector of democracy and leader of the free world). The beginnings of this counter-narrative were evident as early as 1981 (in *Phase 1*) when the Black Caucus wrote a letter to Reagan to express its concern about the signal that the military exercise, "Ocean Venture 81," would send. According to the caucus, maintaining the exercise would be damaging to America's reputation because the exercise would contribute to the militarization of the area and therefore damage America's proclaimed international role.[16]

Concerns over status-signalling sent by the administration were particularly evident during the invasion (29.3 per cent of documents compared to 22.2 per cent before the news of the coup and 10 per cent after the Marines had landed in Grenada). The counter-narrative developed force after news of the invasion was known, and it was constructed around two principal ideas. The first argued that invading Grenada had gone against international law, and, if the US violated international law, it would no longer be respected. Walter Mondale wondered, "How can we justify the invasion under international law?"[17] while Congressman Savage accused Reagan's administration of interfering in the domestic affairs of a sovereign state: "Our nation has no right, legal or moral [even if US citizens are in danger], to dictate by military might the internal affairs of any sovereign nation."[18]

The second idea argued that intervening in Grenada would irredeemably damage or even destroy the legitimacy of the US as leader of the free world. If the US did not respect international law and showed that not respecting it had advantages, why should other countries respect it themselves? Mondale believed the invasion would provide a dangerous precedent for other countries to do the same. It would undermine "the high ground this nation needs as a leader in the world in discouraging other nations such as the Soviet Union from doing what they are doing in Afghanistan, Poland, Cambodia, and elsewhere."[19] Indeed, the argument portraying Reagan as hypocritical in light of criticisms of the Soviet intervention in Afghanistan were common.[20] On the other hand, it was also argued that solving the Grenada conflict through military force, without using this as a last resort, would set up a dangerous practice and precedent for US foreign policy. What guarantee would there be that it would not happen again in the future? After all, new practices begin from exceptions, as noted by Congressman Paul Tsongas: "The concern we have is not the issue of Grenada. It's whether we are going to engage in interventionism. Whether we're gonna be spread throughout the world ... Where does it stop?"[21]

All these concerns harbored fears of losing international support: "The French expressed surprise, the British government has strong reservations. The Canadians are withholding judgment."[22] Such fears, as expressed by TV presenter Ted Koppel on ABC Nightline, were not an epiphenomenon. To find greater nuance in the narratives, I applied three codes to those discourses perceiving the issue as an opportunity to either gain or lose support: feeling unsupported by others, risk of losing support internationally, and feeling supported by others. It is hardly surprising, then, that the first two codes had the highest frequency in public documents, both at 7.1 per cent (compared to 4.9 per cent and 3 per cent for Reagan, secretaries of state, and advisors; and just 5 per cent and 3.5 per cent in the dataset overall). Clearly, public discourses on the invasion were genuinely concerned with the consequences of a hasty or inappropriate status signal, or more generally about the negative implications of the invasion on US status perceptions globally.

Invading Grenada to Demonstrate Great Power Credibility and Resolve? No Real Proof, but Instead a Clear Consciousness of Having to Build a Strong Narrative to Avoid Criticism Damaging to US Status

According to Reagan and his supporters, the US was on its way to losing its great power status because of Carter's terrible presidency. On this point the literature is unanimous: "Few administrations have been elected with a clearer pledge to restore US credibility and greatness abroad than the Reagan administration."[23] For some commentators, Carter's presidency was such a disaster that "many Americans were questioning America's morality and resolve in the world." This corresponds to the views expressed when his presidency is mentioned in the declassified documents. A good example of how Reagan's administration viewed Carter's legacy can be found in a memo sent by McFarlane to William Clark on 4 February 1983:

> The international situation we inherited when we came into office was a truly miserable one. America was the target of countless political and terrorist actions all around the globe. We had lost too many friends in the world. It had become difficult for us to renew military base agreements even with countries that were basically friendly to us and even more difficult to reach new agreements with other countries. It had simply become too politically risky for many world leaders to be closely associated with the United States ... Perhaps the symbolic epitome of this dismal situation was the spectacle of Iranian terrorists seizing American hostages and humiliating them and our country for an entire year.[24]

To put an end to this situation it was vital to send appropriate status signals to prove US resolve and its capabilities as a great power.

The narrative used by the US administration always oscillated between two imperatives: (1) to send correct and appropriate status signals; and (2) to keep a relatively low profile while avoiding criticisms of sabre-rattling. The first issue linked to the idea of sending appropriate status signals (*Phase 1*) was to assure Caribbean countries of US support against Soviet and Cuban expansionism. The policy recommendation provided by the Inter-Agency Group to the NSC on 11 June 1983 is a good example of the dilemma: "It is essential that the US signal its support for the democratic Eastern Caribbean States. A public statement … would carry significant psychological impact."[25] The idea behind keeping a relatively low profile was to avoid making Bishop a martyr to US imperialism.

After news of the coup and Bishop's execution was known, I expected to find more markers of the necessity to send appropriate status signals. This was not the case. In fact, the frequency decreased from *Phase 1* (11.7 per cent) to *Phase 2* (7.8 per cent) then increased in *Phase 3* (9.3 per cent).[26] Even more surprising, a qualitative reading of the dataset revealed that, in the available documents, the status-signalling narrative was a defensive, rather offensive one. By offensive, I mean that such a narrative would have been constructed around the need to intervene in order to save great power status. Instead, the narrative was defensive, constructed around the idea of balancing the narratives developed by critics of the invasion. Only one notable exception to this narrative was found, in the White House talking points of 26 October 1983: "On the international scene, recognition that America accepts its responsibilities and can be counted on to respond to violence firmly, if required to do so."[27]

Otherwise, the narrative was built around arguments to avoid losing support. The US administration was well aware of the criticism that the invasion was establishing a dangerous precedent and it therefore developed a communication strategy to reassure the various audiences, insisting on the uniqueness of the event that shouldn't repeat itself elsewhere.[28] The administration was aware that their narrative had to be carefully built to avoid comparison with the Soviet invasion of Afghanistan.[29]

To achieve this aim, they frequently reminded everyone that if they intervened it was because they had received an urgent call for help from weaker countries – and that helping the weak was part of the US's international role.

Conclusions on Status-Signalling Framework

My analysis revealed the presence of a quantitatively strong status-oriented framework. The situation was often described in terms of status and the conflict was seen as an expression of wider, more global struggles. The understanding that status is relational and gained (or lost) through a process of recognition was also identified; however, its relevance to explaining the US decision to invade Grenada is less certain.

Status appears to have been more so at stake after, rather than during, the time when the decision to invade was made. In summary then, we may conclude that status-related questions clearly played an important role in the Grenada conflict, but more as a supporting framework for holistic interpretation of the events rather than an explanatory one offering strong insights into the decision to invade. Accordingly, when we compare the results for the realist-strategic framework to those for the status-signalling framework, we see that even with a lower quantitative frequency the former presents much stronger empirical evidence to explain the US decision to invade. In other words, realist-strategic considerations were likely much more influential than status considerations on the decision to invade Grenada.

Evaluating the Nostalgic Virility Framework: Ronald is Super Reagan

A Quantitatively Dominant Framework

Explicit and implicit ideas related to the nostalgic virility framework were found in a considerable 41 per cent of documents in the dataset. This overall score makes it the dominant framework of the four tested. Table 8.1 offers a synthetic view of the overall frequency obtained for each of the tested theoretical frameworks.

The relative importance of the nostalgic virility framework was expected based on how the literature describes Reagan and his style on the one hand, and based on feminist literature studying US-Caribbean relations on the other. To draw preliminary conclusions and hypotheses, I looked at the frequency distribution of the framework based on various variables. There was a strong increase in the frequency of the framework once news of the coup and Bishop's execution were known (from 34.7 per cent in *Phase 1* to 48.3 per cent in *Phase 2*). Later, there was a slight decrease, yet the framework remained important in *Phase 3* (46.8 per cent).

Preliminary analysis of the frequency distributions based on actors reveals that evidence for nostalgic virility was found most for Reagan (59.1 per cent), followed by the public (48.6 per cent) and secretaries of state (46.5 per cent). This indicates that Reagan was a nostalgic virile individual, although questions remain as to how these qualities were

Table 8.1 Frequency of the four theoretical frameworks in the Grenada case

Framework	*Frequency (per cent)*
Realist-strategic	18.3
Diversionary theory of war	1.8
Status and symbolic	26.6
Nostalgic virility	41.0

manifested and used, and how his virile understanding of great power status was constructed. As the public appears to have been sensitive or responsive to nostalgic virility, the "bread and blood" narrative also potentially becomes relevant. To verify this hypothesis, we must determine whether US domestic opinion was in need of reassurance about its country's virility. Based on the frequency distribution of evidence for nostalgic virility among the type of documents, we may provisionally confirm the hypothesis. Indeed, the framework appears mostly in public documents (57.8 per cent) and external documents (42.9 per cent), while occurring in internal documents at a frequency that is still important, yet much lower (35.7 per cent).

Based on these initial quantitates, we may conclude that the nostalgic virility framework had a strong presence in the Grenada conflict and that its influence could be vital to understanding the US administration's motives in ordering the invasion. To further test this hypothesis and determine the overall place and importance of nostalgic virility in the decision to invade Grenada, the remainder of this chapter will investigate the findings through a qualitative reading of the dataset.

Emotions and Virility: From Annoyance to Disgust to Urgency to Pride: An Evolution of Emotions following the Coup

Emotions are a key component of nostalgic virility. They accentuate the desires that nostalgic-virile leaders may have to take drastic actions, such as punishment of others for their unworthy or deviant behaviours. They can also strengthen leaders' link to the history of their country by making them view any change in their country's reputation as intolerable. So, did emotions play a role in the Grenada conflict? The large increase in evidence for nostalgic virile discourses between *Phases 1* and *2* outlined above could indicate the presence of emotions of revolt against the coup and its consequences (e.g., civilian casualties), as well as emotions of anger, or even fear, for the US students. Similarly, it could indicate the presence of feelings of revenge, seeking to punish those who perpetrated the coup.

Codes were applied to tag a range of emotions (found in 19.8 per cent of documents) present in the dataset. Emotions were found with increasing frequency as the situation evolved and until it culminated with the coup and brutal murder of Bishop. In *Phase 1*, they are evident in very few documents (4.8 per cent), but this rises significantly to 34.2 per cent in *Phase 2* when news of the coup became known. After the invasion (*Phase 3*), the frequency decreases to 23.4 per cent. These initial results suggest the importance of two different qualitative analyses. The first is chronological and aims to examine whether there was an evolution in the composition of emotions – if the same emotions were present from phase to phase or if new ones appeared while others disappeared. Following from this, the second is to investigate whether or not the rise of emotions was linked directly to the coup and Bishop's execution.

In *Phase 1*, little evidence of emotions was found. When present, they appear in the form of feelings of annoyance linked to two situations during a one-month interval at the beginning of 1983. The first was found in one diplomatic cable exchanged between the US Embassy in Bridgetown and the Secretary of State, where diplomats expressed their annoyance and fatigue at Grenada's behaviour, which they deemed disloyal: "Playing on American respect for the sanctity of private discussions between government officials … the PRG has fabricated a history of United States dealings with Grenada designed to cast the USG as the perpetual villain."[30] The second, linked to the realist-strategic framework, was a feeling of annoyance in the administration[31] over the possibility of appearing weak due to the expansion of communism in the Caribbean, a region considered by the US to be its own backyard.[32] US emotions did not seemingly grow beyond this slight annoyance until Bishop was liberated by his supporters and later executed by the more radical regime on 19 October 1983. Before this, emotions were too low in both quantitative and qualitative terms to suggest that they may have influenced and pushed the US government to decide to invade. The coup, however, changed everything.[33]

In *Phases 2* and *3*, US emotions took a more dramatic turn. Feelings of annoyance disappeared and gave way to sentiments of disgust and horror, which in turn led to a sense of urgency. These feelings were found in a large variety of documents: diplomatic cables, memos, public speeches, preparation notes for meetings, etc. However, as transcripts of key meetings during which the decision to invade was taken are still classified, it is not possible to know whether these feelings were "truly" shared by US decision-makers or present at key discussions during which the decision was taken. Publicly, the events in Grenada were described as "a brutal coup of leftist thugs."[34] Emotions of disgust and horror were then used by Reagan in his address to congress (most likely to gain support), insisting that the complete chaos that reigned in Grenada was creating an explosive and dangerous situation: "The only indication of authority was an announcement that a barbaric shoot-to-kill curfew was in effect."[35]

This feeling of disgust may well have been genuine as it was also found in the administration's internal documents and not aimed to be made public. Two days after the execution of Bishop, the US Embassy in Kingston reported that the situation was worsening because of brutal and humiliating executions: "A woman, Miss Jacqueline Creft, who was minister of education was reportedly beaten to death, the Foreign Minister, Mr Unisson Whiteman was severely beaten before he was executed."[36] Emotions of disgust and outrage could have also influenced the decision to go to war in the context of the Caribbean countries asking the US for help. According to diplomatic reports, leaders of these countries reacted emotionally to the news of the coup: "They [leaders of Barbados, Dominica, St-Vincent, St Lucia] are unanimously horrified and outraged at the barbarity of those who have seized Grenada."[37] However, as key transcripts are still classified and only the official legal document is available for research, it is not known for sure whether emotions played a strong part in their appeals to the US.

This apparently horrific situation created the second main emotion found in the dataset, beginning in *Phase 2*: a feeling of urgency. St George's medical school was attended by hundreds of American students and the chaotic situation resulting from the coup might put them in danger.[38] The origins of this feeling of urgency come not only from the situation in Grenada but also the fact that the US did not see the coup coming: "What, however, surprised us was the intensity of the barbarity by which the change of leadership in Grenada was accomplished."[39] Added to this was the perceived feeling that a decision must be taken in a matter of minutes, as Reagan tried to justify to Thatcher in a phone call on 26 October 1983.[40] So, was it really a perceived sense of urgency or just claimed as one? It is not possible, at this point, to give a definitive answer because transcripts of key meetings are still classified. What is known, however, is that Thatcher was furious not to have been warned in advance of the invasion, despite Grenada being part of the Commonwealth. A sense of urgency (as well as a fear of internal leaks) could well have been an excuse to decrease Thatcher's irritation. However, these remain speculations.

The feeling of urgency was reinforced by analogies to a similar situation that had occurred in 1979: the Iranian hostage crisis, during which fifty-six US diplomats were detained for 444 days. The Carter administration tried a rescue operation that failed miserably. Hundreds of US citizens in Grenada could well have been seen as an opportune target for a government more radical and hostile to America. As such, Mark Adkin argues that "Reagan was haunted by the fear of another Teheran-style hostage situation, with the United States powerless in the hands of a small group of fanatics."[41] Reference to the Iranian crisis was also found in the dataset: "American lives are at stake in the turmoil in Grenada – 500 to 1000 of them. Their safety is paramount. They could be taken hostage, as in Iran."[42] However, it is not known whether or not this analogy and fear was widely shared in the administration or by Reagan himself, as it was identified only a few times in the available documents.

The overarching portrait of US emotions during the conflict would be incomplete without looking at those present in *Phase 3*. While the emotions identified in *Phase 2* continued, a new emotion found exclusively in public documents arose: pride. It was used to describe two very different actors. The first were the US armed forces (identified explicitly as the hero of the invasion in 4.8 per cent of cases), praised for their "gallantry and heroism"[43] not only in the successful invasion but also because of the positive image of the US they sent through their behaviour: "Without exception, local officials praised the manner in which American forces were conducting themselves. Prime Minister Adams [of Barbados] emphasized that there had not been a single incident."[44]

The second actor associated with the emotion of pride can be found in correspondence between the rescued students (and their families) and Reagan. In one such instance, Reagan responds to a letter from a Mr and Mrs Bernard, in which they expressed their support of his decision to invade, despite their children being in potential danger as students in Grenada: "The fabric of our society is kept strong because of families like yours who love and care for another. You are cut from the same cloth as those other families whose young serve our nation and the cause of freedom around the world."[45]

As such, we see that emotions were both present in various forms and considerably important in certain US narratives and rhetorical strategies at the time when the decision to invade was taken. The important place given (whether consciously or unconsciously) to emotional appeals in these discourses illustrates claims that US politics is very theatrical, with US presidents required to perform in front of various national and international audiences – indeed for some, such need for performance can be explained by a previous loss of US virility.[46] Given such claims and my findings on the important role of emotions in this conflict, I want to explore this notion of performativity to see whether it helps un in interpreting the evidence of nostalgic virility identified in the dataset.

A Surprising Presence of Nostalgic Virility: Reagan as a Performing President-Actor

Scholars such as John Ducat and Cynthia Weber have highlighted the importance of virility in US politics. Ducat argues that a key aspect of contemporary US domestic politics and foreign policies is male anxiety.[47] Although his study concentrates on US presidencies from Bill Clinton to George W. Bush, the idea of male anxiety could well have played a key role in earlier foreign policy decisions, such as the invasion of Grenada. Weber argues that if the US invaded Grenada, then it was an attempt to recover the phallus it had lost during the Vietnam War – Grenada was simply a way of signalling the recovery of the phallus.[48] Several elements of the nostalgic virility framework will help us examine whether these arguments are relevant and as important as claimed in the Grenada conflict.

Showing and Reassuring US Domestic Opinion That Uncle Sam Has a Big Stick: A "Bread and Blood" Narrative

Nostalgic virile individuals are those who place great importance on the idea of strength, either at a personal level (e.g., their body) or a symbolic level (e.g., the military). As there would be no strength without a fit body, nostalgic individuals also give importance to the idea of health, again either personally (e.g., being in good physical shape, etc.) or symbolically (e.g., their country being fit because of institutional or moral superiority). These concerns have consequences: nostalgic virile individuals are very sensible to the notion of "losing face," being humiliated, or appearing weak. When confronted with challenges, they tend to be (or appear) firm and resolute. They express a high-level need for power and have a low level of complexity, often making them view issues and solutions as dichotomous. They tend to be nostalgic of a glorious past that they believe has collapsed and that they want to bring back to life. So, based on the data analyzed, were any US actors or decision-makers in the Grenada conflict qualified as nostalgic virile, and if so, how was this used and/or expressed in decision-making and debate?

Quantitative analysis reveals that all the main elements of virility mentioned above are present in the overall dataset. They were found in 16.1 per cent of documents, with emphasis especially on being brave (7.8 per cent), being strong (7.3 per cent), and being firm and resolute (6.8 per cent), reinforced also by explicit mention that invading Grenada would be a sign of courage (3 per cent). They were identified in Reagan's documents and interventions (being brave: 11.4 per cent; strength: 9.1 per cent; being firm and resolute: 15.9 per cent) as well as among the US public (being brave: 23.4 per cent; strength: 22.9 per cent; being firm and resolute: 15.7 per cent). Finally, evidence linked to nostalgia was found in only 4.5 per cent of documents. Before considering any qualitative importance or insights on this low frequency of nostalgic claims, I will outline how the virile aspects of nostalgic virility played out in the Grenada conflict.

As these key aspects of virility were present in Reagan's documents, I expected the qualitative analysis to reveal strong empirical evidence of nostalgic virility as a key argumentative or decision-making logic. In fact, it was the complete opposite. Evidence of virility was found in Reagan's documents, but not in ways that described his understanding of the situation or what was at stake. Instead, it appeared in his documents after US troops landed in Grenada, present mainly in public discourses describing the invasion and the decision-making process that led to it. Moreover, this evidence was found only in Reagan's public documents (letters, speeches, etc.). Following these results, one explanation for this may be that Ducat and Weber's arguments are correct: discourses and actions were first and foremost directed, and probably designed, toward pleasing a national audience and reassuring Americans that there was no US loss of virility.

Qualitative analysis revealed that the idea of strength was present in the form of references to moral strength, and was used to describe the worth of US soldiers, especially those who were injured. In itself, this association would contradict the theory: injuries do not necessarily prove strength. However, the idea of being injured was used to reinforce the myth of the Marine-as-superman, giving him a sort of a mystical aspect. The best example is in Reagan's address to the nation: "It says more than any of us could hope to say about the gallantry and heroism of these young men who serve … I'll let General Kelley's words describe the incident. He spoke of a 'young marine with more tubes going in and out of his body than I have ever seen in one body … He was making signals and we realized he wanted to tell me something. We put a pad of paper in his hand and he wrote *semper fi*.'"[49] Other key virile aspects are rarely found in Reagan's documents other than indicators of dichotomous and clear-cut understandings (both 6.8 per cent) when he explained that there was no other choice but to order the invasion: it was either invading or letting chaos reign. Other dichotomous and binary combinations included the choice between showing strength or appearing weak, and that the students could be either safe or dead.[50] Because of the dichotomy between these two extremes, Reagan argued (and wrote in many responses to letters from American citizens): "The decision to authorize participation by our troops in the rescue mission to Grenada was one of the most difficult of my administration."[51]

These examples do not, however, tell us much, and without further data it would be preferable to adopt a cautious approach to the question of whether Reagan qualifies as a nostalgic virile individual: either he was and the data is insufficient to affirm it, or he was not and simply played at being one – just as an actor plays a role, the actor Ronald plays President Reagan. The idea of Reagan playing a role was well developed by Weber: "One could never be sure if the film actor turned politician who succeeded in capturing the presidency was being himself or playing a role from one of his B movies."[52]

Let us assume that Reagan was "just playing" at being virile. Ideas of strength, being brave, and being firm and resolute were found in 26.3 per cent, 21.1 per cent, and 36.8 per cent of his public documents, respectively, but were wholly absent from external documents and internal documents. The fact that nostalgic virility was present only in public documents could indicate that Reagan either wanted to appear virile in front of the US public or that he believed that appearing as such was what the US public wanted. If this were the case, we would be in a situation recalling the practice of evergetism – destroying an opponent to reassure domestic opinion about the virility of the country, showing that the traumatic events of 1979 could be put behind America once and for all. Another surprising fact in support of this is that virile indicators were associated to Reagan, but in discourses from other actors in which Reagan was the main actor of two opposite virile stories.

A Virility-Oriented Hero-Villain Narrative to Describe Reagan: Californian Rambo or Cro-Magnon Bulldozer?

Reactions to Reagan's actions and discourses were very different depending on political allegiances. This section will outline some of the contours of the US public debate over Grenada to better understand how criticisms and support for Reagan may have shaped or influenced his administration's decision-making and response to the conflict. Qualitative analysis revealed the presence of a hero-villain narrative in which Reagan played both roles (hero in 10.6 per cent of cases; villain in 4.5 per cent). This narrative had a clear virility component, as key aspects of nostalgic virility were used to cast Reagan as either hero or villain. In the dataset, the villain narrative was used exclusively by Democrats and the press, both unsupportive of Reagan, and was based on the ideas of an inappropriate use of force, a lack of sophisticated thinking, and unhelpful nostalgia that prevented him from understanding clearly that the world was changing. It can be synthesized by the image of Cro-Magnon man first using his mace, hitting as hard as he could, and only having discussions afterwards. In this way, Reagan was described as not taking the right decisions because of counterproductive (if not useless) nostalgia: "Reagan is now trapped in the presidential prison, and like presidents before him, he is being buried alive under an avalanche of expectations … in an America which no longer exists or exists only in his mind."[53] His mindset was not the only part of Reagan described as "old": his body was, too, as he was "exhausted by the Beirut tragedy and suddenly made old."[54]

Added to this, he was criticized for having a lack of complexity in his way of thinking (clear-cut understanding and a dichotomous manner of seeing events), which made it harder for him to understand how the world worked: "I wonder if he isn't finding, however, that the world is more complicated than he thought it would be. He is not a naive man, by any means, but I think he's finding that his simple, straightforward views simply aren't working."[55] Virility was seen by Democrats as negative because they believed diplomatic means were always preferable. Congressman Don Barker claimed that Reagan's administration "is using military force instead of foreign policy. It is relying on bullets instead of brains." Congressman Thomas Downay regretted that "this is an administration that shoots first and asks questions later."[56] Reagan's preference for the use of a direct and virile approach to foreign policy was judged catastrophic by Democrats: "We have a trigger-happy president who has gotten us in two civil wars – Lebanon and Grenada."[57]

On the other side, virility was seen as the key component of a heroic narrative starring Reagan himself. Here, every aspect of nostalgic virility was used in a positive way to construct the narrative. Reagan was described as strong and capable of communicating his strength to America, which was no longer taken for a weak country. His resolve and preference for resorting to the use of force to solve the Grenada issue was seen as him being able to solve every issue on the international stage, to such an extent that he would have been able to prevent the Iranian hostage crisis from happening had he been in power at that time: "I always voted Republican but allow me to dream in the past with the Iranian crisis. If you would have been president during President Carter's years the Ayatollah Khomini would have been buried and dead a long time ago because you would have given the same military order … and the Iranian experience would have never come to happen."[58]

If, in the previous narrative, his lack of complexity was seen negatively, it was not so here. Reagan was praised for following his intuition and for accepting risks associated with the use of force rather than waiting to see if there was a real danger: "Your instinct is wonderfully sure. You did exactly the right thing in Grenada. You gave a new way of hope to the people."[59] His display of strength through the use of military force saw him described as decisive and resolute. In the face of danger, he was a man – a "real one" who faced danger and was not a "wimp" who ran away and hoped for the best.[60] He was also seen as someone able to put aside emotions, even in the most dramatic events, in order to concentrate on current issues: "Reagan hasn't been lighting candles and hoping for the best. In the midst of the reporting of the Beirut carnage, he spent Sunday reviewing plans for the Grenada invasion. He conferred throughout Monday on the military moves required."[61]

Finally, he was seen as brave and courageous precisely because the decision to invade Grenada was a risky one not to be taken lightly. Peter Schweizer, in his study of Reagan's struggle against communism, reports that during the decision-making process leading to the invasion, "One mentioned that there would be political fallout, the invasion would be controversial and probably very unpopular. 'I understand that. I am prepared for that' Reagan said."[62] In the available dataset, no explicit mention of being ready to risk his political future and popularity to save students potentially in danger was found. There is evidence of the idea that Reagan took a courageous decision, but it was expressed less explicitly, essentially in letters of support sent by parents of rescued students expressing their gratitude: "Your actions in Grenada served to protect lives of very special Americans. I thank you for your concern and courage."[63]

These opposing discourses highlight that US virility and the possibilities for political action deriving from and upholding it were central to the public debate over the pro and cons of intervening in Grenada. On one side, virility was seen as a disadvantage for the future of US diplomacy, while on the other it was depicted as the correct and only sensible course of action. While these public debates do not show direct evidence of how decision-makers such as Reagan and his administration perceived the conflict, they nonetheless inform and help shape the overall ecosystem of ideas and influences over American foreign policy decisions at the time. The next section will continue to investigate these diverging views about virility by outlining how, in the case of the US invasion of Grenada, virility was constructed around two ideal-types of envisioned response that are, according to George Lakoff, generally mutually exclusive opposites: the strict father and the nurturant parent mindsets.

US Invasion: Between Punishing and Nurturing: A Two-Faced Virility

According to the two ideal-type mindsets of the strict father and the nurturant parent, if the decision to invade was based on a strict father mindset, then we would expect to find references to the idea that Grenada and/or its leaders had behaved badly and needed to be punished by Reagan. If, however, the invasion had been guided by a nurturant parent mindset, we would expect to find references to the idea that the objective was to help other countries (the Caribbean states) achieve their goals. We should also find a lexical field of a positive power relationship between the US and these countries.

As Reagan was a conservative, I expected to find a strict father-oriented narrative. This did occur in 24.1 per cent of documents, but the nurturant parent narrative was also found in 12.3 per cent of documents. How were these two narratives constructed? What are their connections? And what do they reveal about the understanding of the Grenada issue? The following two sections attempt to provide answers to these questions.

Punishing "Fake" Virility and Morally Deviant Individuals: A Strict Father Perspective

To gain a better idea of how the strict father narrative was used, and thus detect possible consequences, I looked in detail at its distribution by narrowing the frequency computation based on the variables stage, actor, and type of document. Overall, the results for each variable were relatively high.

In the evolution of this narrative, it was clear that the coup in Grenada played a role. It was present in 17.1 per cent of documents in *Phase 1*, rose to 32.9 per cent in *Phase 2*, and remained in heavy use after the US forces landed in Grenada (27.3 per cent in *Phase 3*). This is consistent with the theory's predictions: that the events in Grenada caused outrage and, therefore, those who committed and ordered them had to be severely punished. Among actors, the US administration used the narrative the most (33.1 per cent of documents), followed by Reagan (27.3 per cent) and then the public (22.9 per cent). These results are not surprising and confirm Lakoff's thesis that strict father mindsets were mostly found among conservatives (e.g., Reagan and his administration). The narrative was also found far more in public documents (34.9 per cent) than in internal documents (23.1 per cent) or external documents (10.7 per cent).

How was the strict father narrative constructed? Overall, it was composed of the following ideas: being mature and disciplined (14.8 per cent), punishing others (5.5 per cent), and providing protection against evil (7.3 per cent). At first (before the coup), the idea of being mature and disciplined was the most evident (15.4 per cent of documents), whereas the idea of punishing was found in only 2.7 per cent of documents and that of protection against evil in 0.7 per cent.

The US self-identifies as mature and disciplined, which means that its leaders have the capacity to differentiate and put aside emotions to concentrate on what is the "right thing" to do in the national interest. It also follows strict rules. In *Phase 1*, the US administration demonstrated this by not taking hasty decisions regarding Bishop's regime. Despite recurrent provocations and hostile maneuvers, the US repeated many times that it would not take the bait and invade. Instead, it was determined to give him the opportunity to explain his behaviour: "Last June and July we agreed to give Bishop a chance to respond to our major concerns."[64] Such maturity indicates that the US was ready for any type of situation. When information came that the situation in Grenada was heating up (the day before the coup), the Joint Chief of Staff George Schultz was tasked "to draw up more detailed contingency plans for emergency evacuation under hostile and non-hostile conditions,"[65] should they have to evacuate US students.

After the coup, the decision was taken to divert ships to the area as a precaution: "The President decided to divert some naval ships in the area … That was essentially precautionary so that if the situation became worse, we would have a capability nearby."[66] Once US troops landed in Grenada (end of *Phase 2* and beginning of *Phase 3*), the argument about being mature and disciplined took a different form. Discipline was linked to scrupulous respect for the law – both national and international – and especially the War Power Resolution, which fixes the conditions under which the president can use military force: "In accordance with my desire that congress is fully informed on this matter, and consistent with the War Power Resolution, I am hereby providing a report."[67]

During the military operations, the decision was taken to provide detailed information to Cuba and the Soviet Union to assure the safety of their nationals and thus avoid any friction: "At the moment of our landing, we communicated with the governments of Cuba and the Soviet Union and told them we would offer shelter and security to their people."[68] The last expression of US discipline and maturity is found after the invasion. Instead of leaving Grenada and showing no further interest in the country, the US decided to provide help for its reconstruction.

McFarlane detailed US objectives to congressional leaders: "[To] assist the government in preparing free elections. [To] offer technical and financial assistance to create the structures and procedures necessary for a fair and representative electoral process."[69]

The punishment aspect of the strict father narrative began to be used frequently starting in *Phase 2* (10.1 per cent, and dropping to 3.9 per cent in *Phase 3*). In addition, the idea of punishing was reinforced by explicit mention of having the duty of protecting the weak (in 18.1 per cent of cases overall). The invasion was thus viewed as necessary to protect the weak, who were themselves clearly identified in 19.1 per cent of documents: first and foremost were the US students studying at St George's medical school,[70] followed by the Grenadian citizens themselves. These civilian victims are clearly identified in the hero-villain-victim narrative as needing to be saved by US invasion, while the villains of the narrative were, interestingly, not countries or regimes but specific individuals, described in 9.3 per cent of documents as morally deviant and/or as having a "fake virility." In fact, three groups of villains were seen as needing to be punished because of their moral perversion and for faking their virility: Bishop, the New Jewel Movement leaders, and Fidel Castro.

Bishop (and members of his government) was obviously the first villain to be identified. Three grievances against him were identified. First, accusations of moral perversion because of his ostentatious support of communism: "Bishop, unfortunately, is not interested in UFOs [a reference to the previous Prime Minister, Eric Gairy, who Bishop deposed in 1979 and who spent his time looking for UFOs], just Cubans and Russians."[71] Second was his deviant behaviour, namely for putting his own citizens in danger and doing nothing to improve the situation, not even asking for external help. A cable sent from the US Embassy in Bridgetown to the secretary of state the day after Bishop was put under house arrest summarizes the situation in Grenada very clearly: "It is also very evident that there are in fact life-threatening situations in Grenada. The earlier reports of lack of electrical power and attendant loss of refrigerated medical supplies. There have been no official requests."[72] Such grievances against Bishop were not new. Third, US officials judged him untrustworthy and as someone who was always reconsidering his decision of whether or not to have talks with the US. Despite his various claims that he wanted to open such discussions, he repeatedly took a hostile position against the country by supporting every Cuban and Soviet decision.[73]

The second group of deviant actors to be "punished" were the leaders of the New Jewel Movement who had planned the coup and were seen as responsible for the bloodshed that followed. They were described as "sadistic pro-Cuban and anti-US"[74] and as fanatics jeopardizing the safety of innocent civilians: "Their [members of the Organization of Eastern Caribbean Sates] security is directly and adversely affected by the violent and bloodshed deeds of the most radical faction of the New Jewel Movement."[75] With such "qualities," it's no wonder Reagan was said to be very unwilling to trust the movement.[76]

The final villain identified was Castro. His depravity was denounced when it was known that he had ordered Cuban nationals in Grenada to fight to the death against US troops – despite the numerous US messages in which its officials informed Castro of their intentions and targets, and despite the slim chance they had to successfully stand against the US and thus end the conflict. Lives were sacrificed for nothing: "Cuban resistance continues. Colonel Tortola acknowledged Castro's message of 26 October urging the Cubans to fight to the last man."[77]

Was US domestic opinion receptive to this strict father narrative? It appears so. Overall, 86 per cent of the population declared approval of Reagan's special address to the nation discussing Grenada, and of his actions in response more generally (compared to 64 per cent before the speech).[78] More specifically, the core of Reagan's electorate seemed very receptive to it and seemed to be asking for more. A letter sent to Reagan from a supportive Republican on 27 October 1983 eloquently states: "Mr President, as a plain American and a redneck and as a man that loves America the Great and the beautiful, I respectfully, Mr President, would like to suggest the following ... when you've finished your undertaking in Grenada, I respectfully request Mr President that you send the USS Independence [one of the flagships of the task force sent to Grenada] and more up to Nicaragua, El Salvador and any other area of disturbance to our way of life."[79]

Assisting and Supporting the Weak: A Nurturant Parent Perspective

Indicators of the nurturant parent mindset were found in 12.3 per cent of cases overall. Just like the strict father narrative, it was frequently used beginning in *Phase 2* (18.4 per cent compared to 6.8 per cent in *Phase 1*). In a striking difference from the strict father narrative, the nurturant parent narrative was used primarily by Reagan (31.8 per cent) and not at all by the US public (0 per cent). The other main difference is that the nurturant parent narrative was found mostly in external documents (28.6 per cent compared to 7.3 per cent for public documents and 13 per cent for internal documents).

Before moving on to an analysis of how the nurturant parent narrative was built, a few words are necessary when considering these two differences. Due to lack of data, what follows can only be considered speculative, but I do believe it to be of interest nonetheless. I propose that the absence of the nurturant parent framework within the public discourse can be explained by the practice of evergetism discussed in chapter 8: the American public did not strongly feel like it was expressly helping others (i.e., Grenada) and was more attached to the idea of displaying strength (and thus, punishment). Furthermore, the surprising (perhaps counter-intuitive) finding that Reagan embodied this nurturant parent narrative to a fairly strong degree may be explained, in part, by the fact that it's found mostly in *external documents*, due probably to the need to reassure other countries about US intentions and to avoid charges of imperialism. In this respect, the nurturant parent narrative, built around ideas of supporting and assisting weaker actors, would seem less frightening or bellicose than the strict father narrative, which was built around the idea of punishing and a direct call for the use of military force.

The nurturant parent narrative was built around providing assistance to weaker actors though a positive power relationship and has three major expressions, depending on which stage of the dispute is examined. All three narratives, although different in their expression, share one foundation: Caribbean countries demonstrate goodwill but lack both the means and capabilities that their ambition requires, and thus need the help of the US who can show them what to do. In *Phase 1*, it was essential for the US to provide a platform for Caribbean countries to help their economic and democratic development, hence the Caribbean Basin Initiative. As Reagan wrote to the Mexican President de la Madrid: "It is one of the major goals of my administration to assist the countries of that region to find peace and strengthen democracy."[80]

Once the coup had occurred, and an official request for help had been sent by the Organisation of Eastern Caribbean States to the US, the nurturant parent narrative evolved. It was first a question of the US supporting the Caribbean countries by helping them to get the broadest support from as many other countries as possible. As Reagan told

Pierre Charles, the Prime Minister of Dominica: "The United States will support your efforts in the OAS to obtain the broadest possible endorsement."[81] Furthermore, the participation of US military forces was judged as absolutely necessary because of the weakness and lack of experience of the Caribbean countries' forces. The US Embassy in Bridgetown provided an assessment in a cable to the Secretary of State: "They [Barbados and Jamaica] are untrained, ill-equipped and have virtually no experience with the kind of operation they contemplate. The others have only domestic police forces."[82] After the invasion, the nurturant parent narrative took the form of helping Caribbean countries build a new democratic regime in Grenada.[83]

At first sight, the fact that both the strict father narrative and the nurturant parent narratives were found together – notably both within Reagan's own discourses – could simply indicate a further divide in US public discourse and decision-making regarding virility and the possibilities for potentially contradictory actions. However, based on the qualitative analysis it appears that both frameworks acted together in a complementary manner to serve the same objective: showcasing US superiority and thus continued virility.

Conclusions on Nostalgic Virility

From the frequency of use of nostalgic virile arguments and qualitative analysis of the dataset, it is possible to affirm that this framework is important and provides the most solid explanation for the US decision to invade Grenada. The need to punish deviant actors is clearly evident, and there seems to be an important emotional link between the decision to invade and the way the coup occurred. Evidence for a second virile narrative also appeared – that of evergetism – which took the shape of invading Grenada and destroying the opposition in a brutal display of force to demonstrate to the American public that their country was, and continued to be, the most powerful. However, given the current documents available and a lack of more rigorous or systematic evidence, this second narrative and its purported influence on the decision to invade Grenada can remain only as hypothesis. Surprisingly, the nostalgic aspect of the nostalgic virile framework was quite absent from the data – a surprising finding given the fact that Reagan centred his first presidential campaign around the need to end US decline.

General Conclusion: Which Framework Best Explains the US Invasion of Grenada?

Before undertaking the analysis of each theoretical framework, I provided some a priori evidence in support of each framework tested. The coding frequencies and qualitative analysis of the results allowed me to validate some frameworks and invalidate others, while in other cases it was not possible to offer a conclusive answer for or against hypotheses given limitations of data or weak results. These analyses have also enabled new hypotheses to be put forward which must be tested once further data is available for research.

My analysis began with a study of the mainstream theoretical frameworks of realism and DTW (chapter 8) before proceeding to an examination of my proposed framework of status-signalling and nostalgic virility in this chapter. The mainstream frameworks obtained mixed and ambiguous results. The realist-strategic framework consists of a vast selection of theories and hypotheses involving the pursuit of material and geopolitical interests. It was tested first because it was the most logical or intuitive one in the context of the Cold War. Quantitative analysis shows this framework to be significant, which was also confirmed by qualitative analysis unpacking important narratives surrounding a Cold War-oriented hero-villain dynamic. The idea of going to war to contain and roll back the spread of communism was also identified, although questions remain as to whether this narrative was the trigger for the US invasion. What raises doubts is that the framework seems to be more of a post-fact narrative to justify the invasion and gain (or avoid losing) public and international support. If the framework was the key element, we would have found it at a much more frequent rate in *Phase 2*, and with stronger qualitative expressions.

The second mainstream theoretical framework tested was the DTW. This predicts that wars happen when decision-makers fear losing power in the short term. To them, going to war provides the perfect opportunity to appear as capable leaders; they also count on a rallying-around-the-flag effect to divert opinion from domestic issues. Quantitively, this framework was very weak; qualitatively, it initially appeared completely irrelevant. However, during analysis of status and the nostalgic virility frameworks, I reconsidered my position. It is true that there is not much direct evidence in favour of the DTW framework, but – as many status and virile arguments were used in the public documents, and given the fact that the US public was very receptive to these arguments – it is possible that the DTW framework played a more important role than originally envisaged. If this is the case, then the DTW framework would be present and constructed around an argument as follows: that the US administration went to war in order to reassure American domestic opinion about the virility (military strength) of their country and that it remained, therefore, the same powerful country it used to be. Nevertheless, this virile DTW must remain a hypothesis until access to more classified documents can be obtained for further testing.

Finally, I then tested my proposed theoretical frameworks of status-signalling and nostalgic virility, constructed around the nostalgic-virile understanding of great power status held by some decision-makers. I argue that nostalgic virility makes leaders very careful about status and appearing strong, as well as ensuring that challenges (direct or symbolic) do not go unanswered. This framework seemed to be the most robust explanation (both quantitatively and qualitatively) of the decision-making logics and arguments that triggered the US invasion of Grenada. Key elements were identified, such as the need to send appropriate status signals, important symbolic issues being at stake, and the need to punish morally deviant villains (and those "faking" their virility).

With how much confidence can these claims made? It is true that, although numerous, the documents in the Reagan Library are lacking in diversity, and there are key documents still unavailable for research. Accordingly, the claims made in this chapter cannot be asserted with absolute confidence. Despite this, the dataset has enabled me to invalidate some claims made in the literature, the most important being that Grenada was seen as an area of disturbance important enough to justify a planned invasion, in which case the consequences of the October coup were only a lame excuse. All the evidence found in my dataset argues against this claim. On the other hand, the major hypotheses I put forward surrounding the modernization and "virilization" of evergetism and the public and political use of virile and symbolic narratives to please domestic opinion must be investigated further.

9

Conclusions and Implications for Studying the Causes of War, China and Russia

Understanding Nostalgic Virility as an Explanatory Model for War

This book considered the question *Why do decision-makers of great powers go to war to respond to a provocation made in their country's periphery by a less powerful actor?* I have aimed to put forward a new model linking status-seeking behaviour to leaders' personal belief systems surrounding nostalgic virility as a way of approaching this question.

The concept of nostalgic virility can be understood as a belief system articulated around a nostalgic and virile understanding of what it means to be a great power. Nostalgic virility can be defined at the very least as a combination of the following traits: (i) an emphasis on physical strength; (ii) an emphasis on traditional virile values (e.g., firmness and bravery); (iii) the use of narratives that forward and enhance ideas of moral and behavioural superiority and set examples and models for others to follow; (iv) a positive understanding of a country's history and legacy, which serves as a reference point to assess and understand situations; and (v) a short time preference (i.e., any issue has to be addressed immediately).

The complete theoretical model (i.e., the interaction of status-seeking behaviour and nostalgic virility traits) argues that when faced with a challenge from a less powerful actor in their periphery, leaders of great powers will react by a powerful military response because of their nostalgic-virile understanding of great power status. The aggressor has not only to be punished, but punished severely, and not really because of the aggression in itself, but mostly because of its symbolic aspects. The consequence of this challenge is that leaders (as well as their country) then appear as weak, uncertain, or "worse," effeminate or castrated and in decline. This image is accentuated by a glorified interpretation of the past. Through their mental representations of what represents great power status, leaders have a tendency to understand any aggression in their periphery by a less powerful actor as an international insult (or even as a narcissistic insult). Such challenges demonstrate that deterrence by great power status did not work, that their country (or even themselves as leader of the country) is no longer respected, nor feared. It is not only the image that others have of their country that is at stake if no immediate and strong action is taken in response to the challenge, but also the image of the leader of a great power themselves that other leaders hold (and that the domestic opinion can also hold) that risks collapse.

Key Findings of This Study

The cases treated here and their affirmative results raise a host of comparative questions about great powers' military interventions in their periphery when they are challenged by a weaker actor. Let's briefly revisit some of the overarching findings for each theoretical framework tested here.

Realist-Strategic Framework

Surprisingly, the realist-strategic framework, probably the most dominant in IR literature, performed poorly in all three cases. If the framework was found at all, its role appeared marginal and accounted for very few of the arguments identified as contributing to the decisions to go to war. Realist scholars would surely have expected material power to have played a key role (either in the form of competition between great powers, or in an attempt to shift the power balance). Scholars following a geopolitical approach would surely have expected the strategic location of Indochina, the Falklands, and Grenada to play a major explanatory role in these conflicts. The table below shows that such expectations did not play the major role expected.

From a quantitative perspective, the French and British cases seemed similar. In the French dataset, strategic aspects did not play any major role, and were essentially used by MPs to criticize the government or to express concerns with the war. The strategic and geopolitical context of the Cold War had, it seems, absolutely no effect on French decisions during the specific time frame analyzed (1945–47), although it was present when the French military briefed the government about the Chinese

Table 9.1 Comparison of the realist-strategic framework in the three case studies

Case	*Frequency of arguments in dataset (per cent)*
Indochina	2.8
Falklands	3.6
Grenada	18.0

position (for obvious reasons). In the British case, the framework seemed important only at one critical moment: when Britain needed to secure US logistical support during the task force transit to the Falklands, which meant securing US agreement to use American military facilities on Ascension Island. Britain hoped to gain support by projecting to the US an undesirable alternative reality in which they would be deprived of British support by their own actions. This strategic and diplomatic employment of the framework's assumptions had its uses, but does not in itself explain the British decision to go to war. When examining the invasion of Grenada, there was also a strategic use of the framework (creating an immediate threat to the US to justify and legitimize the invasion after it was over).

Overall, however, we find that regardless of the use of realist-strategic formulations and assumptions within diplomatic relations and political communications, the framework itself offers little explanatory force for understanding the decision to go to war in these three cases.

The DTW Framework

Qualifying the importance of the DTW to the cases analyzed is a harder task than assessing the importance of the realist-strategic framework. If we rely on statistics alone, then table 9.2 would create the impression that the DTW framework had a relatively important role only in the British-Falklands case.

Qualitative analysis suggests, rather, that the relative importance of DTW-based assumptions in each of the three cases can be ranked as follows.

At the bottom would be the French case: the situation in Indochina did not impassion the French at all and there appears to be no attempt to divert domestic opinion away from internal problems. As regards the

Table 9.2 The quantitative importance of the DTW in the three cases

Case	*Frequency of arguments in dataset (per cent)*
Indochina	0.3
Falklands	13.6
Grenada	1.8

British case, the findings show that, contrary to beliefs in the existing literature that Thatcher invaded the Falklands to save her political career, the DTW framework was present only insofar as it was used (similarly to the realist-strategic one) as a strategic diplomatic narrative for securing vital US support. Finally, the American case is more puzzling. In concrete terms, the DTW framework was explicitly used only by the press and congress to criticize Reagan, who, according to them, fabricated an international crisis to increase his image, demonstrate his statesmanship, and prove he was anti-Carter. Despite finding no evidence for this in any of the Reagan administration's documents, are these criticisms far from the truth? I suggest that there might have been an attempt to use the situation in Grenada to increase Reagan's political standing, based on an association of the realist-strategic framework with the idea of being a virile leader. In short, the hypothesis is that, the strategic threat posed by Grenada was symbolically increased post-invasion to secure Reagan's image in the eyes of Republican voters by showing that he was not afraid of taking drastic action in potentially threatening situations, thus securing his image as a "virile" leader.

Status-Seeking Behaviour Framework

Both the quantitative and qualitative analyses have shown that status and symbolic aspects were extensively discussed in all of the three cases, as table 9.3 recapitulates.

In all cases, the analysis sought to identify the ways in which great power status was understood by decision-makers as a basis for understanding the potential dynamics of nostalgic-virile status-seeking behaviour. The British and American cases are very similar in terms of how great power status and the roles associated with it were understood (protecting democracy, protecting and enforcing international law, and protecting the weak). The French case offered interesting variations.

Table 9.3 Comparing the presence of the status and symbolic framework across the three cases

Case	*Frequency of arguments in dataset (per cent)*
Indochina	33.3
Falklands	17.7
Grenada	26.6

Although military strength and capacity of projection were seen as important, the roles associated to great power status were different. The French associated great power status with the role of being a "lighthouse in the dark," to bring their civilization to remote areas.

What can explain this difference? One hypothesis could be the nature of the political regimes. The US and Britain were both under conservative liberal governments, so it is not surprising that the idea of protecting democracy and international law were seen as important: it constitutes a backbone of liberalism, especially the kind of liberalism conceptualized within IR theories. France, on the other hand, seemed to exemplify the "old world," one that virtually disappeared after the Second World War. Most politicians during the French Fourth Republic were already present in the Third, the regime that really engaged France in colonialism and saw its notion of exporting civilization as a primary justification. An intriguing aspect of the French case is that, despite these old habits, there were a few hints that the characteristics identified in the British and American cases were slowly starting to emerge.

What is clear, however, is that the literature claiming that status has to be signalled and is associated to specific roles that need to be appropriately performed appear to be correct.[1] My results show that leaders had a clear idea of what roles were associated to their country's great power status, and that they should send detailed and appropriate signals to either keep or recover this status among their peers and less powerful challengers. In all three cases, there was a clear need to send appropriate status signals before, during, and after the decision to engage in war. In the three cases it was found, also, that great powers' military interventions in their periphery were a long way from garnering systematic and unanimous international support, even from close allies. France had to face US hostility and struggled to secure British support; the UK worked

Table 9.4 Comparing the presence of nostalgic virility in the three cases

Case	*Frequency of arguments in dataset (per cent)*
Indochina	47.7
Falklands	40.7
Grenada	41.0

hard to secure vital US support; and the US faced strong British criticism during its invasion of Grenada.

Nostalgic Virility Framework

Finally, the three cases also showed that nostalgic virility had a significant impact and was, in fact, the dominant explanatory framework in all three cases.

This claim can be made, despite the internal difference between the three cases, because the same key narratives and aspects of the theoretical framework were found in similar contexts. The first was the presence of the traditional characteristics of virility: strength, bravery, and firmness. Strength was portrayed in all cases in the same way: symbolic physical strength because of a strong military capable of achieving goals even in difficult situations. Bravery was found associated to the military, too, but also to leaders themselves, mostly in their capacity to take difficult and unpopular but necessary decisions. Firmness was associated with leaders and expressed around their capacity to stick to their decisions. Empirical evidence showing a refusal to appear weak or effeminate gives support to the works and findings of Ducat and Weber.[2]

The second key aspect of the nostalgic virility framework supposes that nostalgic-virile individuals have a low level of complexity[3] and that appearing strong is key. These aspects were clearly identified in the British and American cases, although here some caution should be kept as important documents remain classified. The French case proved more difficult to analyze, notably because of the high turnover of individuals in command during the new French regime's infancy, especially at the level of head of government.

The third key aspect was the need to punish what leaders identified as morally deviant behaviours (targeting, especially, "fake" or deviant virility, viewed as the opposite of their own), based on what Lakoff identifies as a strict father mindset.[4] What each country fought against was defined in essentially identical terms: a fake or deviant virility lacked resolve, acted in a cowardly way, was irresponsible, and broke moral and legal (domestic and international) rules. The French and American cases, unlike the British, also displayed evidence of what Lakoff identifies as the nurturant parent mindset – a finding that is not so surprising given the colonial context of the conflict in Indochina and the specific nature of relations between the US and the Caribbean.

These key elements of the nostalgic virility framework were also found in other contexts and situations not originally envisaged, thus increasing the impact and importance of the framework's explanatory power and analytical possibilities. For example, it was found in hero-villain narratives used to criticize leaders, especially Thatcher and Reagan, who were said to lack finesse and be more of a "bulldozer" type of actor. Although less clear in the French case, there are hints of such a narrative at work in criticism of the government and the military.

In all cases, emotional discourse was present, constructed around the same emotions and evolving in the same direction. At first, there was the feeling of humiliation caused by a weaker opponent that had fooled the great power with false intentions. In the French and British cases, this was reinforced by the harsh treatment of their military; at a later stage, these negative emotions were replaced by more positive ones, mostly feelings of pride due to military successes.

Evidence in the three cases shows that virility also indirectly shaped debates. In the French case, there was an implicit debate about how to understand virility on several occasions, notably over how to respond to events in Haiphong. Opposition between those wanting to strike back and those wanting to appease does not seem to have been based on the absence/presence of virility, but rather on competing conceptions of what virility really means. In the British case, some documents indicate that the policy disagreements between Thatcher and some of her ministers (e.g., Carrington and Ridley) over policy options had their roots in the debate between the presence/absence of nostalgic virility that shaped the way Thatcher understood and defined British status. The US case revealed the presence of the same debate, and virility was criticized for making decision-makers misunderstand when to respond to a given situation.

Implications for Studying Great Power Politics and IR: Is Nostalgic Virility Relevant to Broader, Contemporary Contexts Such as Ukraine and the South China Sea?

What are the implications of these results for the study of IR beyond the three specific case studies from France, Britain, and the US? Could other theoretical positions held in the IR field benefit from the findings of this research? The analysis tested the suitability of several frameworks to explain why great power decision-makers went to war in their country's periphery against a less powerful actor. Though the scope of this research was constrained to cases in which the outcome was war, it is worth considering how it might contribute more broadly to the study of IR-related phenomena.

Two main sets of implications arise from this study: (1) theoretical implications for the field of IR theory; and (2) empirical-methodological implications for the study of contemporary conflicts. I will conclude by briefly discussing each in turn.

Implications for IR Theory

How might this research contribute to the understanding of diversionary wars? The assessment of the DTW model in the three case studies turned out to be fairly unsuccessful. However, the British and American cases provided some material for the study of diversionary wars, mostly by suggesting the need for a switch of research interest from leaders' personal political concerns (which were not found in any of the three cases) towards domestic opinion, human passions, and emotions. In both cases, I found evidence suggesting that the decision to go to war could have been taken to sway domestic opinion. Due to a lack of data, this point remains hypothetical and needs further testing, but it seemingly consists of a modernization of the old Roman practice of evergetism, which would link a leader's political fate to status and virility. In such a situation, and depending on how the ideas of strength, bravery, and national status are understood by public opinion, leaders wanting to remain in power would be implicitly forced to choose to go to war due to a sort of "bread and blood" symbolic dynamic among the populace.

This research thus provides strong empirical results showing that virility (understood as a system of beliefs around the ideas of strength, resolve, bravery, morality, and health) can be an important factor in decisions to go to war, and in this regard, that it is not only leaders who have ideas (consciously or not) about what it means for their state to be virile. Just as individuals seem to analyze situations and crises under a virile spectrum, the very idea of virility also shapes general understandings and oppositions about how to conceive of roles on the international stage, about what power means, and about how it should be used.

These results, of course, support the claims of several scholars that the study of individuals is pertinent to IR. My findings thus engage positively with studies arguing that individuals matter and that it is therefore relevant to study them and not simply stick to the use of abstract objects (such as the state). In this sense – and perhaps more importantly – my results also engage synergistically with various approaches to studying leaders,[5] helping, I believe, to bridge these differing theories. Nostalgic virility goes beyond individual characteristics like "punitive/nurturing mentalities," "high/low level of complexity," "hawkish behaviour," and "reputation crusader," and incorporates them into one unified theory that offers a broader investigation of how individuals (self-)reflect on key concepts of IR, such as power, justice, status, and roles – all of which come into play at some point during the decision-making process.

Finally, the empirical results obtained in this study can contribute to existing IR debates. One such ongoing debate is about the understanding of status. One camp argues that it is granted through material capabilities and that status is a tool to advance objectives.[6] The other camp argues that status is relational and supposes a certain type of behaviour to follow.[7] From the analysis of the three datasets studied here, I found more evidence in favour of the second camp. This is not to deny the legitimacy of the first, but rather to simply point out that evidence was very strong showing that individuals (ranging from domestic opinion to decision-makers) all had a fairly clear idea about what great power status meant, and that status talks turned not so much on material aspects but were essentially centred on which status signals to send, what it meant to have great power status, and how to continue to fulfil (self-)attributed roles.

The results of this study could, therefore, contribute to the further operationalization of power transition theory (PTT) models. Greve and Levy have already taken a step forward in linking revisionist positions to status dissatisfaction by introducing the "individual" factor in the PTT model – i.e., looking for status dissatisfaction in public speeches.[8] Despite this turnaround, they continue to sit in the middle of two obvious positions: examining the speeches of decision-makers for indicators of status dissatisfaction on the one hand, while on the other adopting an anthropomorphist view of the state. Instead, the model this book proposes combines the individual and national levels of analysis without over-personalizing states while investigating individual actors' emotions and perceptions and states' public decisions. Namely, linking virility to the perception of having reached power parity would then be a more versatile and accurate tool than simply understanding power parity from the purely material point of view of most realist scholars. Virility would likely act to push leaders towards the decision of launching a revisionist policy and engaging in war.

Implications for Empirical-Methodological Approach to Studying Contemporary Conflicts: The South China Sea and Ukraine[9]

I'd like to conclude this study by pointing towards some of the ways in which the framework of nostalgic virility can be relevant and applicable to diverse, contemporary conflicts. This study theorizes that there is an ideal of status-seeking, nostalgic-virile leaders – a combination of behaviours and belief systems that should be understood as probabilistic rather than deterministic. Nostalgic virility is not present in exactly the same manners, proportions, or nuances across the three cases tested, and it would certainly not be present in identical ways in other cases. The conflicts considered here were cases belonging to recent history, but what about more current situations?

The model developed in this book argues that it is probable that most individuals will not fall neatly into the nostalgic-virile category and will not match all the criteria theorized therein, but that those who do are likely to display signs and tendencies that push them towards deciding to go to war when challenged. Recent events suggest that we would likely find status considerations combined with nostalgic virility in contemporary power distribution contexts. For instance, current situations which seem to be implicated in dynamics of nostalgic-virile status seeking include Xi Jinping's positions regarding the South China Sea and Vladimir Putin's war against Ukraine.

Of course, these examples have not all led to full-blown war, but if we look at their respective discourses and associated policy positions with the status-seeking and nostalgic virility model in mind, we would intuitively find markers of a nostalgic-virile understanding of status combined with aggressive foreign policies and aggressive rhetoric. As such, in order to test the relevance of the nostalgic virility model to contemporary cases, the exact same coding dictionary used for this study has been applied to Russian discourses about Ukraine and Chinese discourses about the South China Sea. Of course, these studies are limited in the sense that only public documents are available and not declassified ones, as with the cases investigated in the book. Nonetheless, the tendencies identified give good indication of the ongoing relevance of nostalgic virility and demonstrate several similarities between the Russian and Chinese cases and those of Indochina, the Falklands, and Grenada. For example: both Chinese and Russian officials use a hero-villain-victim narrative in which they are positioned as the hero and the US as the enemy; and both use history as a political weapon to legitimize their actions while presenting themselves as "mature and disciplined" in contrast to the "immature" and "deviant" US, thus leading to the need to engage in strong punitive behaviour. Let's consider these trends in slightly more detail to better understand the contemporary relevance and manifestations of nostalgic virility on the global stage.

Xi Jinping and China's Policy towards the South China Sea

The South China Sea (SCS) issue really captured headlines in 2014 after the Philippines put the issue of territorial rights in front of the Permanent Arbitration Court, a move which infuriated China. The issue concerns territorial reclamation in the South China Sea and is mostly contested by China, Vietnam, and the Philippines. The SCS is a good example for studying Chinese foreign policy because it crystalizes status issues, strategic interests, and international legal issues together into an historically and culturally complex conflict. Based on analysis of the available data, Chinese officials use the same kind of narratives as those explored in this book: nostalgic virility elements are found in 42.3 per cent of documents overall, while purely status-related considerations are found in 19.2 per cent and realist-strategic arguments in just 7 per cent.

The starting point of the Chinese position is found within a perceived feeling of humiliation and disrespect (identified in 9.1 per cent of documents). This feeling is not new and can be traced back to the "century of humiliation," but it is seemingly expressed more violently now that China has the material means to seek revenge. The fact that the US and occidental countries can interpret and stretch certain international legal norms to their interests, but that China cannot, is seen as a deeply humiliating order and positioning for the Chinese state: "so why can't China exercise its lawful right."[10]

This feeling of humiliation is used to frame the issue within the broader spectrum of a fight between a hero (China) that needs to punish a villain (the US and its allies) – a narrative which can be found in 28.8 per cent of cases. China is the hero of the story and the upholder of a "true" virility not only because its actions in the SCS are aimed towards "harmony" and the good of the international community,[11] but also because China claims to have an orthodox understanding and practice of international law. Chinese officials claim to be strictly respecting international law and to not be increasing tensions despite being victimized by other countries.[12] With this argument, Chinese officials indicate that while many countries (especially the US) would have reacted violently, China, on the contrary, behaves in a peaceful and civilized way. In parallel, nostalgic virility is used in another narrative to strengthen Chinese legitimacy and diminish that of the US, based on the principles of role theory, especially of alter-casting combined with Todd Hall's "emotional diplomacy" to criticize the US (13.3 per cent of cases overall). The US, this argument goes, is not just the villain in the SCS, but it is a world villain because it manipulates the content of international law to sustain its hegemonic geopolitics.[13]

As a consequence, the US and its allies have to be punished because they represent a deprived virility (10.1 per cent) by demonstrating immature behaviour and creating instability in the region.[14] As such, Chinese officials develop the idea that neighbouring countries need to receive orders from China in order to allow them to be(come) civilized. In this manner, Chinese discourses have taken a more authoritarian turn: sentences become constructed around action verbs in their negative forms ("don't," "stop," etc.) and the use of the present participle tense, to which indicators of immediate action are added (e.g., "now," "urge," etc.). If such immature behaviours continue, then "the Chinese side will definitely react" and "take firm actions in response" to punish the aggressor.

In Chinese discourses, virility is associated to a specific understanding and use of history that is identified in 22 per cent of documents. History is politicized and used in several ways:

(1) It fuels virility by implicitly recalling past times when China was humiliated by Japan and occidental countries, and by portraying these past times as coming to an end and therefore supporting the need to seek revenge. This state of mind motivates the systematic recourse to History in Chinese legal discourse, as if it were a matter of both avoiding new humiliations and denouncing those of the past. These threats are identified based on countries' past behaviour. For example, the West is considered untrustworthy because of the "century of humiliation," and Japan is described as a major threat because of what it did during the Second World War: "We hope that Japan can learn from history, stick to the path of peaceful development, and act prudently in military and security policy."[15]

(2) Then, history is also given another role in China's nostalgic-virile discourses: to support China's claims and legitimacy to propose an alternative world order. In its discourses, Chinese elites aim to merge historical legitimacy with legal legitimacy. To achieve such a fusion, its legal discourse uses the principles set out in international jurisprudence on land and maritime territorial disputes, focusing on the concept of historical rights. These discourses rely empirically on archeology and on recalling events of a glorious past, such as the expeditions of the fifteenth-century Ming admiral Zheng He, to establish that it has dominated the entire region since the Han Dynasty (11.3 per cent).[16] In the Chinese discourse, the use of History is sufficient in itself to establish Chinese rights. This is illustrated by the various semantic constructions used by Chinese officials as they talk of "historical facts," "historic rights" (and not "historical"), "historical practices," "historical evidence," and "historical truth." By resorting to a reinterpretation of the past to demonstrate the existence of its present-day rights, the PRC is instrumentalizing history in the service of a power-based policy. This practice is reinforced by Chinese legal scholars' actions, who select only favourable sources, ignoring contradictory sources, such as the fact that the region is marked by perpetual

contestation of China's positions by all bordering states. As both international jurisprudence and the arbitral tribunal reject this view, Chinese discourses contain a double critique. First, that international law as currently interpreted is a tool of colonization, thus evoking the TWAIL literature.[17] Second, that actual international organizations are biased, corrupted, and rigged (5.2 per cent).[18]

(3) It supports a specific geopolitical representation of China that calls for the establishment of a sort of Chinese Monroe Doctrine, hidden behind a regionalized view of the international order: only countries forming part of a region would be legitimate in dealing with issues of the region. This position is also intended to be applied to other regions and conflicts such as the Syrian crisis[19] and the Yemeni civil war.[20] This position seems to recall the idea of *Großraum* developed by Carl Schmitt, which consists in an international order composed of regional orders, each managed by a great power and having its specific working principles and institutions.[21]

Through this nostalgic-virile discourse, China sees itself as the natural power in this area, the origin of which is to be found in the traditional Chinese system of tribute, which placed China at the top of a natural hierarchy allowing the maintenance of a regional order. This systematic recourse to a nostalgic-virile discourse suggests that China is seeking to avoid further humiliation while denouncing those of the past and expelling the US from the area.

Russia and the War in Ukraine

In February 2022, Vladimir Putin announced that Russia would conduct, in his words, "a special military operation" in Ukraine, thus bringing back to life high-intensity warfare – something that for many seemed to belong to a past era. Since then, the war in Ukraine has already had many repercussions across the world. It is not my objective here to comment on the normative aspects of the conflict, but simply to provide some findings from a rigorous quantitative and qualitative analysis of key Russian officials' speeches about Ukraine since 2012, and to investigate the relevance of nostalgic virility for understanding an ongoing high-intensity war such as this.

So, without going into thick detail, what discursive trends emerge in official Russian discourses on the subject? As with previous cases studied in this book, realist-strategic motivations are present in Russian discourses about Ukraine (11.8 per cent of documents), but they account for considerably less argumentative force than status considerations (21.7 per cent) and nostalgic virility (63.8 per cent). In addition, emotions are often used by Russian officials (49.3 per cent) in support of their status and nostalgic-virile narratives. So, based on this high-level quantitative overview, how are Russian narratives constructed and used, and what do they reveal about Russian elites' view of the world?

Like for China, the starting point in Russian discourses is found within a hero-villain-victim narrative in which the villain is the US (8.7 per cent) and its allies (essentially NATO supported by Ukraine since 2014) (43.5 per cent),[22] the hero is Russia and its armed forces (7.2 per cent), and the victims are Russia (17.4 per cent) and the population living in Donbass (33.3 per cent). This hero-villain-victim narrative combines realist-strategic considerations with status considerations and feelings of humiliation that create the need to seek revenge and serves as the basis for a strong nostalgic-virile discourse. For the period 2008–14, the main villain trying to victimize Russia is the US, because "they immediately tried [after the collapse of the USSR] to put the final squeeze on us, finish us off, and utterly destroy us."[23] This villain is then helped in its task by additional villains – namely NATO supported by Ukraine against its will: "many Ukrainian airfields are located not far from our border. NATO tactical aviation deployed there … will be capable of striking at our territory."[24] Russia is also victimized by the US because of a perceived policy of humiliation and disrespect towards Russian status: "it is unacceptable to talk to a country which is a permanent member of the UN Security Council in such a manner."[25] The other victim of the story is the Donbass population, namely because of the cultural genocide they have suffered since 2014[26] and their regular bombardment by Ukrainian artillery,[27] which is, according to Russian officials, a current practice of the West – a claim for which they regularly use History and historical analogies as supporting evidence (17.4 per cent), in this case placing the Donbass population as "successors" of the Western-led suffering of Serbians, Iraqis, and Syrians.[28] The ultimate proof that Russia is victimized, and that the Donbass population is a victim, is found in the fact that Russian institutions are depreciated while European countries describe Ukraine as a democracy despite its link with neo-Nazi groups: "The European countries' praise of democratic values in Ukraine is puzzling at a time when official Kiev and various Nazi organisations created with its tacit approval, and often direct support, glorify the atrocities of the SS during WWII when they wiped out entire families in Eastern Europe. Importantly, in 2016, the Polish parliament qualified the crimes of the 14th SS Division 'Galicia' against the Polish population as genocide."[29] Putting an end to these actions requires villains to be severely punished.

The need to punish the Ukrainian government, the US, and more broadly Western countries is found in 49.3 per cent of documents. This punitive discourse is built around the ideas that the villains are incapable of change (20.3 per cent) and must have guidelines of proper conduct imposed on them (11.6 per cent) by a mature and disciplined country – i.e., Russia (42 per cent). Not only are the villains portrayed as immature and unable to improve their behaviour on their own, but they are also a symbol of a deviant virility (21.7 per cent), or even a "castrated" virility depending on what discourse we look at. The description of Ukraine and the US as deviant actors takes multiple shapes. They are deviant, first, because their leaders are deviant from a biopolitical perspective (physically weak and unfit to lead). For example, Dmitri Medvedev described Barack Obama as being prone to "mental aberration,"[30] and later Sergei Lavrov implied that European leaders were taking decisions under the influence of alcohol and on outdated data,[31] while Vladimir Putin criticized Western leaders using a vocabulary that recalls the semantic field of deviant sexual practices.[32] These deviant characteristics led them to adopt deviant behaviours on the battlefield and in politics, such as developing biological weapons,[33] using forbidden weapons,[34] and supporting neo-Nazi groups to remain in power.[35]

By contrast, Russia is framed as legitimate in its military operations because it is a mature and disciplined country (42 per cent). This self-description is based on several perspectives. The first one is that Russia scrupulously follows international law and respects its international obligations despite unfavourable contexts. Before 2022, the examples given by Russian officials were the respect of contractual obligations to deliver gas[36] and the claim that Russia was looking for a peaceful and diplomatic solution to the issues in Ukraine.[37] After 2022, the example often given is that Russian soldiers followed strict rules of engagement and targeted only military objectives (by opposition to the Ukrainian military targeting civilians),[38] thus fulfilling Russia's self-described role on the international stage: protecting and promoting regional peace (13 per cent). In addition, Russian officials argue that if they intervene militarily, it is because they no longer have the choice. This is found in discourses through the use of words indicating a dichotomous and clear-cut way of thinking (5.8 per cent). It is, as the discourse goes, impossible to back off, "it is a matter of life and death,"[39] and there is no alternative to the objectives announced by Russia,[40] thus decreasing the likelihood of finding diplomatic solutions to the conflict.

The final important element identified in Russian discourse about Ukraine, and which supports the need for the military operation – which is in fact a punitive operation – is the political use of History in a similar manner to what Chinese officials are doing regarding the SCS issue. Namely, Russian officials frame the current situation in Ukraine as a repeat of past events (13 per cent). The historical situation often presented, and which is seen as impossible to allow to happen again (8.7 per cent), is the Barbarossa operation of 1942.[41] Russian officials clearly describe the current war as a continuation of Nazi invasions, thus requiring Russia to react in the exact same way in order to reproduce the exact same result, i.e. total victory: "your fathers, grandfathers and great grandfathers did not fight the Nazi occupiers and did not defend our common motherland to allow today's neo-Nazis to seize power in Ukraine."[42] Based on the logic of nostalgic virility, there is only one option for Russian officials: "it is simply no longer possible to leave crimes of this kind without retaliation."[43]

Conclusions on Nostalgic Virility in the Decision to Go to War, Both Past and Contemporary

These results indicate that evidence of the characteristics of nostalgic virility is present in different situations and contemporary contexts. The preliminary study of Russian and Chinese discourses about Ukraine and the South China Sea confirms the continued importance of nostalgic virility. Although the study of these two situations could only be based on public discourses, it is clear that nostalgic virility structures Russian and Chinese discourses in several complex ways. It matters either because Russian and Chinese officials believe that a nostalgic-virile discourse will appeal to their domestic audience, or because nostalgic virility really is a component part of their thinking and understanding of the world. In either case, Western countries should take this finding into account and rethink and adjust their diplomatic approaches to incorporate this element and all that it entails. One such approach could include developing a narrative that takes into account the importance of nostalgic virility in order to help decrease tensions and avoid Chinese and Russian officials perceiving the situation as a deadlock in which only aggressive measures offer a way out. More broadly, we may conclude that to correctly understand certain countries' positions on the international stage, we should pay greater attention to emotions and subjective interpretations of history and status, rather than simply over-exaggerating material-strategic and internal political factors.

Another important takeaway that these results suggest is, quite simply, that the model of nostalgic virility has a strong potential for uncovering broader theoretical insights given that, in the three cases analyzed here, nostalgic virility was found not only to be associated with other frameworks (DTW, realist-strategic framework), but also to be a structuring theme around which every other framework interacted and developed. Accordingly, across different contexts, different time periods, and different analytical and theoretical foci, nostalgic virility appears to form complex synergies within discursive and ideational fields, and thus to play an important role in decision-making on foreign (and perhaps also domestic) affairs.

NOTES

Introduction

1 Gilpin, *War and Change in World Politics*; Friedberg, *The Weary Titans*; Volgy, Corbetta, Grant, and Baird, *Major Powers and the Quest for Status*; Mearsheimer, *The Tragedy of Great Power Politics*; Allison, *Destined for War*; Greve and Levy, "Power Transition, Status Dissatisfaction and War"; Shifrinson, *Falling Giants*; MacDonald and Parent, *Twilight of the Titans*; Murray, *The Structure for Recognition in International Relations*.

2 Mearsheimer, *The Tragedy of Great Power Politics*; Hamilton-Hart, *Hard Interests, Soft Illusions*; Raine and Le Mière, "Southeast Asia: Between Emerging Great Power Rivalry"; Kaplan, *Asia's Cauldron*; Navarro, *Crouching Tiger*; Hayton, *The South China Sea*; Allison, *Destined for War*; Mérand, *Coping with Geopolitical Decline*.

3 Mueller, *Retreat from Doomsday*; Keegan, *A History of Warfare*; Holsti, *The State, War, and the State of War*; van Creveld, "The Waning of Major War"; Henchan and Vasquez, "The Changing Probability of Interstate War."

4 Ramel and Holeindre, *La Fin des Guerres Majeures ?*

5 Deriglazova, *Great Powers, Small Wars*.

6 Mack, "Why Big Nations Lose Small Wars"; Jervis, *Perception and Misperception*; Fischerkeller, "David versus Goliath"; Bobić, *Why Minor Powers Risk Wars*; Avey, *Tempting Fate*.

7 Hanink, *How to Think about War*, xlviii.

8 Apart from: Geiss, *Geschichte im Überblick*; Finnemore, *The Purpose of Intervention*; Taliaferro, "Power Politics and the Balance of Risks."

9 Finnemore, *The Purpose of Intervention*, 2–3.

10 The ways in which we understand nostalgic virility and how it is actually embodied may therefore differ between individuals and depend on the historical contexts in which they operate.

11 Waltz, *Theory of International Politics*; Walt, *The Origins of Alliance* and *Revolution and War*; Mearsheimer, *Tragedy of Great Power Politics*; Shifrinson, *Falling Giants*.

12 Friedberg, *The Weary Titans*; Snyder, *Myths of Empire*; Rose, "Neo-Classical Realism and Theories of Foreign Policy"; Zakaria, *From Wealth to Power*; Ikenberry, *Power, Order and Change.*

13 Greve and Levy, "Power Transition, Status Dissatisfaction, and War."

14 Volgy et al., *Major Powers and the Quest for Status.*

15 Wohlforth, "Honor as Interest in Russian Decision for War"; Huberman, Loch, and Onculer, "Status as a Valued Resource"; Plourde, "The Origins of Prestige"; Volgy et al., *Major Powers and the Quest for Status*; Paul, Welch Larson, and Wohlforth, *Status in World Politics.*

16 Markey, "The Prestige Motive in International Relations"; Lebow, *Why Nations Fight*; Murray, *The Structure for Recognition in International Politics*; Wolf, "Respect and Disrespect in International Politics"; McCourt, *Britain and World Power*; Heller, "Russia's Quest for Respect"; "Defending Social Status"; "Russia's Power Politics towards Ukraine."

17 Jervis, *The Logic of Images in International Relations*; Suzuki, "Seeking 'Legitimate' Great Power Status in Post-Cold War International Society"; Lindemann, *Causes of War*; Pu and Schweller, "After Unipolarity"; Paul et al., *Status in World Politics*; Fikenscher, Jaschob, and Wolf, "Seeking Status Recognition through Military Symbols"; Renshon, *Fighting for Status.*

18 Lindemann, *Causes of War*; Ringmar and Lindemann, *The International Politics of Recognition*; Murray, *The Structure for Recognition in International Politics.*

19 Lindemann, *Causes of War.*

20 Chiozza and Choi, "Guess Who Did What"; Horowitz, Stam, and Ellis, *Why Leaders Fight.*

21 Post, *The Psychological Assessment of Political Leaders.*

22 Levy, "The Diversionary Theory of War"; Richards, Morgan, Wilson, Schwebach, and Young, "Good Times, Bad Times, and the Diversionary Use of Force"; Smith, "International Crisis and Domestic Politics"; Tir and Jasinski, "Domestic-Level Diversionary Theory of War"; Tir, "Territorial Diversion: Diversionary Theory of War and Territorial Dispute'; Fravel, "The Limits of Diversion."

23 Popper, *Conjonctures and Refutations*; Przeworski, Alvarez, Cheibub, and Limongi, *Democracy and Development*; Chiozza and Goemans, *Leaders and International Conflicts.*

24 Chiozza and Goemans, *Leaders and International Conflicts*, 19.

25 Levy, "Prospect Theory, Rational Choice and International Relations"; Davis, *Threats and Promises.*

26 Taliaferro, "Power Politics and the Balance of Risk."

27 Hermann, "Assessing Foreign Policy Role Orientation"; "How Decision Units Shape Foreign Policy"; "Assessing Leadership Styles"; Hudson, "Birth Order of World Leaders"; Dyson, *The Blair Identity*; Horowitz et al., *Why Leaders Fight*; Yahri-Milo, *Who Fights for Reputation.*

28 Mueller, *Retreat from Doomsday*.
29 Crawford, "The Passion of World Politics."
30 Jervis, *Perception and Misperception*; Janis, *Groupthink*.
31 Davis, *Threats and Promises*; Lebow, *Why Nations Fight*.
32 Bowman, *Honor: A History*.
33 Crawford, "The Passion of World Politics"; Callahan, "National Insecurities"; Ahmed, *The Cultural Politics of Emotions*.
34 Fierke, "Agents of Death"; *Political Self-Sacrifice*; Clément, Lindemann, and Sangar, "The 'Hero-Protector' Narrative."
35 Weber, *Faking It*; Nussbaum, *Upheavals of Thought*; Ducat, *The Wimp Factor*; Saurette, "You Dissin Me?"; Wolf, "Respect and Disrespect in International Politics"; Brice, "Un president en colère"; Barnhart, "Prestige, Humiliation and International Politics."
36 Hall, *Emotional Diplomacy*.
37 While the theoretical study of the importance of status has begun to produce a rich vein of literature (Lebow, *Why Nations Fight*; Wohlforth, "Unipolarity, Status Competition and Great Power War"; Volgy et al., *Major Powers and the Quest for Status*; Pu and Schweller, "After Unipolarity"; Barnhart, "Prestige, Humiliation and International Politics"; Onea, "Between Dominance and Decline"; Fikenscher et al., "Seeking Status Recognition through Military Symbols"; Ward, *Status and the Challenge of Rising Power*), this research has mostly been conducted through the angle of individuals (Markey, "The Prestige Motive in International Relations"; Wolf, "Respect and Disrespect in International Politics') or decision-making groups (Wallace, *War and Rank among Nations*; Doran, Hill, and Mladenka, "Threat, Status Disequilibrium, and National Power"; Wohlforth, "Unipolarity, Status Competition and Great Power War"; Onea, "Between Dominance and Decline"; Larson and Schevchenko, "Russia Says No"), and has rarely been associated directly to ideas, individual perceptions, emotions, and virility as factors influencing the choice of whether or not to go to war.
38 The definition of "periphery" used here is found in chapter 1.
39 Barkin, *Realist Constructivism*.
40 Lebow, *Why Nations Fight*.
41 Jervis, *The Logic of Images*.

Chapter One

1 Gochman and Maoz, "Militarized Interstate Dispute"; Volgy et al., *Major Powers and the Quest for Status*; Baron, *Great Power Peace and American Primacy*; Mearsheimer, *The Tragedy of Great Power Politics*.
2 Modelski and Morgan, "Understanding Global War."
3 Schweller, *Deadly Imbalances*.

4 Bull, *The Anarchical Society.*

5 Volgy et al., *Major Powers and the Quest for Status.*

6 Congyan, "New Great Power and International Law"; Duque, "Recognizing International Status."

7 The situation of the UK during the period 1815–50 is a good illustration of this point. In 1815, there were 204,000 personnel in the army, but by 1850 there were less than 100,000; there were 89,000 men in the Royal Navy in 1815, and between 23,000 and 39,000 from 1820–50. Despite these declining numbers, the UK was still seen as the major military power and faced no challenges (Marx, *Histoire de l'Angleterre*, 418).

8 Lukes, *Power a Radical View.*

9 Barnett and Duvall, "Power in International Politics."

10 Pardesi, "Is India a Great Power?"

11 Blechman and Weinland, "Why Coaling Stations Are Necessary in the Nuclear Age"; Harkavy, *Great Power Competition for Overseas Bases.*

12 Granvaud, *Areva en Afrique.*

13 Brown, "Do Great Powers Have Great Responsibilities?"; Congyan, "New Great Power and International Law"; Petiteville and Placidi-Frot, *Négociations internationales*; Forsberg, Heller, and Wolf, "Russia and the Quest for Status."

14 Nolan, "Great Powers and International Society."

15 Morris, "How Great Is Britain?"

16 Jackson, *The Global Covenant.*

17 Evans and Sahnoun, *The Responsibility to Protect.*

18 Mercer, "Rationality and Psychology in International Relations"; Deng, *China's Struggle for Status*; Abdelal, Herrera, Johnston, and McDermott, *Measuring Identity*; Paul et al., *Status in World Politics.*

19 Marshall, *Class, Citizenship and Social Development*; Paul et al., *Status in World Politics*; Forsberg et al., "Russia and the Quest for Status."

20 Jervis, *The Logics of Images; How Statemen Think*; Lebow, *Why Nations Fight.*

21 Scott, *The Birth of a Great Power System*; Paul et al., *Status in World Politics*; de Waal-Andrews, Gregg, and Lammers, "When Status Is Grabbed and When Status Is Granted."

22 Badie, *L'impuissance de la puissance*; Kinne, "Dependent Diplomacy"; Albaret and Placidi-Frot, "Les petits Etats au Conseil de sécurité."

23 Abdelal et al., *Measuring Identity*; Lindemann, *Causes of War*; Jervis, *How Statemen Think.*

24 History is full of examples of countries failing to answer challenges appropriately, resulting in consequences such as opening the door to further challenges – as the Qing dynasty experienced when it encountered Europeans (Ford, *Mind of the Empire*).

25 O'Neill, *Honor, Symbols and War*, 91.

26 Reynaud, *Société, espace et justice.*

27 Galtung, "A Structural Theory of Imperialism"; Wallerstein, *The Modern World-System.*

28 Aklaev, *Etnopoliticheskaia Konflitktologia*; Deriglazova, *Great Powers, Small Wars.*

29 MacMahon, "How Periphery Became the Centre," 21.

30 Taliaferro, "Power Politics and Balance of Risk."

31 Sewell and Ryan, *Foreign Policy at the Periphery*, 8.

32 Taliaferro, "Power Politics and Balance of Risk."

33 Aklaev, *Etnopoliticheskaia Konflitktologia*; O'Neill, *Honor, Symbols and War.*

34 Geiss, *Geschichte im Überblick.*

35 Tuckner, *The Purposes of American Power*; O'Neill, *Honor, Symbols and War*; Kertzer, *Resolve in International Politics.*

36 Corbin, Courtine, and Vigarello, *A History of Virility*, 44–5.

37 Bell, *Men on Horseback.*

38 Corbin et al., *A History of Virility*, 51.

39 Cosme, *Auguste.*

40 Applegate, *The Egyptian Book of Life.*

41 Bonhême and Forgeau, *Pharaon : les secrets du pouvoir.*

42 Kantorowicz, *The King's Two Bodies*; Perez, *Le Corps du Roi.*

43 Corbin et al., *A History of Virility.*

44 Ibid., 269.

45 Ibid., 249.

46 Wood, "Performing Memory"; Hill and Gaddy, *Mr Putin*; Foxall, "Photographing Vladimir Putin"; Goscilo, *Putin as a Celebrity and Cultural Icon*; Sperling, "Putin's Macho Personality Cult."

47 Foxall, "Photographing Vladimir Putin."

48 Ibid. Similar trends are found in the 2016 US presidential election, wherein a focus on Clinton's physical health became a large focus and frame of coverage after one of the debates where she appeared to be a little unwell. See Glenn Kessler's article published in the Washington Post on 18 August 2016. Article accessible at: https://www.washingtonpost.com/news/fact-checker/wp/2016/08/18/trumps-claim-that-clinton-lacks-the-physical-stamina-to-be-president/.

49 Mosse, *The Image of Man*; Stets and Burk, "Feminity/masculinity"; Goldstein, *War and Gender*; Ducat, *The Wimp Factor*; Connell, *Masculinities*; Shepherd, *Gender Matters in Global Politics*; Sjoberg, *Gendering Global Conflict*; Sperling, "Putin's Macho Personality Cult."

50 Sjoberg, *Gendering Global Conflict.*

51 Moffitt, *The Global Rise of Populism.*

52 To date, there is no dataset measuring traits of female leaders, possibly because "women's entry into political leadership is a relatively new phenomenon, with over half of female world leaders having come to power since 1990 … The relative

paucity of female heads of state makes it statistically impossible to test for effects of gender, making the use of case studies crucial for understanding the motivation of those individuals" (Horowitz et al., *Why Leaders Fight*, 160). In terms of the types of virile emotions women express, studies have shown that there are no real differences from those expressed by men (Shibles, "Feminism and the Cognitive Theory of Emotions"; Parashar "Anger, War and Feminist Storytelling").

53 Ducat, *The Wimp Factor*, 136.

54 Horowitz et al., *Why Leaders Fight*.

55 Campbell, *Margaret Thatcher*; Garnett, "Banality and Politics"; Ponton, "The Female Political Leader."

56 Aron, *Paix et guerre entre les nations* ; Sperling, "Putin's Macho Personality Cult."

57 Carrère d'Encausse, *Six années qui ont changé le monde*.

58 Weber, *Faking It*, 5.

59 Corbin et al., *A History of Virility*.

60 Matard-Bonucci and Milza, *L'homme nouveau dans l'Europe fasciste*.

61 Courtois, *Ernst Nolte*, 655.

62 Dyson, *The Blair Identity*, 29.

63 De Larrinaga and Doucet, "Governmentality, Sovereign Power and Intervention."

64 Lakoff, *The Political Mind: Why You Can't Understand 21st Century*; *Moral Politics*; *The Political Mind: A Cognitive Scientist's Guide*.

65 Khong, *Analogies at War*; Boin, Hart, Stern, and Sundelius, *The Politics of Crisis Management*; Aragones, Gilboa, Postewaite, and Schmeidler, "Rhetoric and Analogies."

66 Boulding, "National Images and International Systems."

67 During the 2012 presidential debates between Mitt Romney and Barack Obama, for example, the former accused the latter of having decreased the capacity of the military, based on the number of naval units in operation (22 October 2012, Barack Obama/Mitt Romney debate, Commission on Presidential Debates, https://www.debates.org/voter-education/debate-transcripts/october-22-2012-the-third-obama-romney-presidential-debate/). Another example: the size of the British Royal Navy. Some argue that it needs more ships as it would not have been able to reconquer the Falklands in its current state (2 April 2017, "Royal Navy Far Weaker than It Was during Falklands War," Laura Hughes, *Telegraph*, article accessible at: https://www.telegraph.co.uk/news/2017/04/02/britains-navy-far-weaker-falklands-could-still-cripple-spain/).

68 Darwin, *The Empire Project*, 612.

69 Aron, *Paix et guerre entre les nations*; O'Neill, *Honor, Symbols and War*; Lebow, *Why Nations Fight*.

70 MacDonald and Parent, *Twilight of the Titans*.

71 Cartledge, *Sparta and Lakonia*; Goldstein, *War and Gender*.

72 Tang, "Reputation, Cult of Reputation, and International Conflict."

73 Allès, "Ethnic and Religious Violence," 364.
74 Baranets, *Comment Perdre une Guerre.*
75 Darwin, *The Empire Project*, 611–12.

Chapter Two

1 Saurette and Gordon, *The Changing Voice of the Anti-Abortion Movement.*
2 Fairclough, *Critical Discourse Analysis*; Wodak and Meyer, *Methods of Critical Discourse Analysis.*
3 Klotz and Prakash, *Qualitative Methods in International Relations.*
4 Reinert, "Mondes lexicaux stabilisés et analyse statistique de discours."
5 Baillat, Emprin, and Ramel, "Des mots et des discours."
6 Halperin and Heath, *Political Research.*
7 Fierke, *Diplomatic Interventions.*
8 Laver Benoit and Garry, "Extracting Policy Positions from Political Texts"; Dyson, *The Blair Identity*; Halperin and Heath, *Political Research.*
9 Saurette, "You Dissin Me?"; Reus-Smit and Snidals, "Reuniting Ethics and Social Science"; Brice, "Un president en colère."
10 Tang, "Reputation, Cult of Reputation, and International Conflict."
11 Bergmane, "La consultation des archives des affairés étrangères."
12 The transfer of government and administration data was systemized and made mandatory by law during the presidency of Valerie Giscard d'Estaing (Law 79-18, 3 January 1979).
13 This study is based on preliminary results. The bigger project that aims at comparing the Russian invasion of Ukraine to the Chinese policy in the South China and how these two decisions challenge the current international order and attempt to make a new world emerge will be presented at two international conferences in 2023: the 27th annual world convention of the Association for the Study of Nationalities, Columbia University; and the conference "Rethinking the Past and Present of Liberal Internationalism" organized by the London School of Economics and the University City of London.

Chapter Three

1 As a reminder, *Phase 1* goes from 24 March 1945 to 18 December 1946 (from de Gaulle's speech in which he highlights French objectives and strategy regarding Indochina to the day before the Viet Minh attack in Hanoi). *Phase 2* goes from 19 December 1946 to 22 March 1947 (from the day of the Viet Minh general attack against the French in Hanoi to the vote of war credits). *Phase 3* goes from 23 March 1947 to 23 December 1947 (the date that any negotiations with the Viet Minh have been *de facto* officially ruled out).

2 Gros, *Histoire de la guerre d'Indochine*; Ferrari and Verret, *Une guerre sans fin*; Pottier, "A French Experience in Counterinsurgency"; O'Dowd, *Military Strategy in the Third Indochina War*.
3 Elgy, *Histoire de la IVe République*.
4 De Wailly, *L'Empire Rompu*.
5 Dalloz, *La guerre d'Indochine*; Ruscio, "L'opinion française et la guerre d'Indochine"; Valette, *La Guerre d'Indochine*; Franchini, *Les Guerres d'Indochine*; Wainstock and Miller, *Indochine et Vietnam*.
6 Paxton and Hessler, *L'Europe au XXième Siècle*.
7 Badie, "French Power-Seeking and Overachievement"; Schmitt, "Decline in Denial: France since 1945."
8 Vallaud, *La Guerre au XXème siècle*.
9 Corbin et al., *A History of Virility*.
10 Grimal, *La décolonisation de 1919 à nos jours* ; Ruscio, "L'opinion française et la guerre d'Indochine"; Franchini, *Les Guerres d'Indochine*.
11 Elgy, *Histoire de la IVe République*, xxiii.
12 Dalloz, *La guerre d'Indochine*; Franchini, *Les Guerres d'Indochine*.
13 Burbank and Cooper, *Empires*.
14 Ibid.
15 Gueniffey and Lentz, *La fin des empires*.
16 Jean Cédile was a former resistance fighter and officer in the Free French Armed Forces during the Second World War, and later a public servant. He was airdropped into Indochina in 1945 and tasked with re-establishing French authority at a time when France was not able to send an expeditionary force. He was later captured by the Japanese. After his release he was tasked to negotiate with Ho Chi Minh.
17 17 September 1945, Leclerc Papers, 239 K 4, Private Archives, reproduced in its entirety in Bodinier, *Le retour de la France en Indochine*, 175–83. This narrative was later used (and supported) by left-wing MPs such as Antoine Dumusois: 23 August 1946, *Officiel de l'Assemblée Nationale*, No. 82, 3276.
18 18 March 1947, *Journal Officiel de l'Assemblée Nationale*, No. 26, 897.
19 Elgy, *Histoire de la IVe République*.
20 13 March 1947, *Journal Officiel de l'Assemblée Nationale*, No. 24, 832.
21 28 May 1945, General Staff of National Defence File, Series 4 Q (77), Military Archives.
22 23 December 1946. *Resistance Paris-matin*.
23 4 December 1946, *Journal Officiel de la République Française*, No. 115, National Assembly; see also, 18 March 1947, *Journal Officiel de la République Française*, No. 26, National Assembly.
24 29 November 1946, F/60/3035, National Archives.
25 Elgy, *Histoire de la IVe République*.

26 25 May 1946, report from Juin, Joint Chief of Staff, after his mission in the Far East, Leclerc Papers, 238 K 6, Private Archives, reproduced in Bodinier, *Le retour de la France en Indochine*, 261–6.

27 5 October 1946, *Journal Officiel de la République Française*, No. 109, National Assembly.

28 14 March 1947, *Journal Officiel de la République Française*, No. 24, National Assembly.

29 8 December 1946, Indochina Files, Series 10 H (165), Military Archives.

30 Soutou, *La Guerre de 50 ans.*

31 23 December 1947, *Journal Officiel de la République Française*, No. 132, Consul of the Republic, 2524.

32 6 January 1947, Files F/60/3035, Interministerial Conferences Folder, National Archives.

33 19 February 1947, Indochina Files, Series 10 H (950), Military Archives.

34 3 May 1947, minutes, Indochina Files, Series 10 H (159), Military Archives.

35 Elgy, *Histoire de la IVe République.*

36 18 March 1947, *Journal Officiel de la République Française*, No. 26, Session No. 30, National Assembly.

37 Ruscio, *La Guerre Française d'Indochine.*

38 Wainstock and Miller, *Indochine et Vietnam.* The French press often reported on illegal financial and speculative practices surrounding the "piastre" (Indochina's currency that was indexed on the Franc). The piastre suffered several devaluations as a consequence of the implementation of the Bretton Woods system. To compensate, France was financing the value difference between the Franc and the Piastre to companies who were conducting businesses based on the Piastre. This led to an increase of practices such as fake bills and contracts. The Viet-Minh used these practices and the inability of the Indochinese financial offices to control the situation to acquire French weapons. Ultimately, the situation was revealed in 1952 by journalist Jacques Despuech.

39 Dalloz, *La guerre d'Indochine*; Ruscio, "L'opinion française et la guerre d'Indochine."

40 Ruscio, "L'opinion française et la guerre d'Indochine."

41 2–3 September 1945, *Le Populaire.*

42 17 September 1945, Leclerc Papers, Series 230 K, Private Archives, reproduced in its entirety in Bodinier, *Le retour de la France en Indochine*, 175–83.

43 8 December 1946, "situation ayant conduit aux opérations commencées le 20 novembre 46 au Tonkin," Bulletin Study 48, Series F/60/3035, Indochina Committee Folder, "brides de données de séance," National Archives.

44 9 May 1947, note from high commissioner, Indochina Files, Series 10 H (167), Military Archives.

45 2 November 1947, report from Valluy to Battet, Indochina Files, Series 10 H (165), Military Archives.

46 General Louis Morlière was commander of military operations aimed to liberate the South of France during the Second World War. He was then sent to Indochina, where he was partisan of a cautious approach with the Viet-Minh, as opposed to Admiral d'Argenlieu, governor of Indochina, and General Valluy, who both were partisan of a punitive approach. The personal oppositions are representative of internal divisions in the government and French military about how to deal with the Viet-Minh. This division was not without creating operational problems (Le Courriard, "Les socialistes et les débuts de la guerre d'Indochine").

47 5 December 1947, Series F/60/3035, Indochina Committee Folder, "brides de données de séance," National Archives.

Chapter Four

1 28 August 1945, report from Leclerc to d'Argenlieu, Chief of Defence Staff Files, Series 4 Q (78), Military Archives.

2 25 May 1946, report from Juin on his mission in the Far East, Leclerc Papers, 238 K 6, Private Archives, reproduced in its entirety in Bodinier, *Le retour de la France en Indochine*, 261–6.

3 E.g., 22 October 1946, Léon Pignon's report, Indochina Files, Series 10 H (143), Military Archives.

4 E.g., in report written from Leclerc to President of the Republic in 1947, Leclerc Papers, 239 K 4, Private Archives.

5 23 December 1946, Paul Bastid, *L'Aurore*.

6 14 March 1947, *Journal Officiel de la République Française*, No. 25, Session No. 29, National Assembly.

7 Franchini, *Les Guerres d'Indochine*; Wainstock and Miller, *Indochine et Vietnam*.

8 Frémeaux, *De quoi fut fait l'empire ?*; Holeindre, "A Certain Idea of Grandeur."

9 18 March 1947, *Journal Officiel de la République Française*, No. 26, Session No. 30, National Assembly.

10 20 October 1945, Indochina Files, Series 10 H (569), Military Archives.

11 27 March 1947, telegram to high commissioner, Indochina Files, Series 10 H (165), Military Archives.

12 De Wailly, *L'Empire Rompu*.

13 5 December 1947, note from Morlière, File F/60/3035, Indochina Committee Folder, "brides de dossiers de séance," National Archives.

14 Wainstock and Miller, *Indochine et Vietnam*.

15 2 October 1945, telegram from de Gaulle, National Defence Staff Files, Series 4 Q (1), Military Archives.

16 What happened in Haiphong and the importance of the incident continues to divide historians on whether to make it the starting point of the Indochina Wars (the starting point being a customs issue). The French navy stopped a Chinese boat smuggling oil, and the Vietnamese opened fire on the French, including civilians. Valluy favoured and gave orders in favour of a powerful response, while others such as Debès were in favour of a more cautious approach.

17 10 December 1946, report on events of 20 November 1946, Indochina Files, Series 10 H (602-2), Military Archives.

18 12 December 1946, report on situation in northern Indochina, Leclerc Papers, 239 K 4, Private Archives, reproduced in its entirety in Bodinier, *Le retour de la France en Indochine*, 365–72.

19 9 May 1947, high commissioner to Bollaert, Indochina Files, Series 10 H (167), Military Archives.

20 17 September 1945, report from Cédile to Leclerc, Leclerc Papers, 239 K 4, Private Archives, reproduced in its entirety in Bodinier, *Le retour de la France en Indochine*, 175–83.

21 25 September 1945, report from Leclerc on situation in Indochina, Leclerc Papers, 239 K 4, Private Archives, reproduced in its entirety in Bodinier, *Le retour de la France en Indochine*, 185–92.

22 27 September 1945, telegram from Leclerc to the Minister of Foreign Affairs, National Defence Staff Files, Series 4 Q (76), Military Archives.

23 De Wailly, *L'Empire Rompu*.

24 30 August 1945, note to Palewski, Files AG3/(4), National Archives.

25 14 February 1946, telegram from Leclerc to Juin, Indochina Files, Series 10 H (143), Military Archives.

26 Title given to prime ministers under the IVth Republic.

27 21 October 1947, Inter-ministerial Conference on situation in Indochina, Files F/60/3035, National Archives.

28 1 February 1947, note from Valluy to reinforcement soldiers, Indochina Files, Series 10 H (297), Military Archives.

29 26 March 1947, *Journal Officiel de la République Française*, No. 26, Session No. 30, National Assembly.

30 Chairman of the National Government of the Republic of China (the "nationalist China" that will later become The Republic of Taiwan, also known as Taiwan, after his defeat to the communists, directed by Mao).

31 28 August 1945, report from Leclerc to d'Argenlieu, Series 4 Q (78), National Defence Staff Files.

32 5 February 1946, note from Valluy to the President of the Republic, Files F/60/3035, Indochina Committee Folder, "brides de données de séances," National Archives.

33 Valette, *La Guerre d'Indochine*.

34 27 February 1946, note from Leclerc to Salan, Indochina Files, Series 10 H (163).

35 25 May 1946, report from Juin on his mission in the Far East, Leclerc Papers, 239 K 6, Private Archives, reproduced in its entirety in Bodinier, *Le retour de la France en Indochine*, 261–6.

36 17 September 1945, report from de Cédile to Leclerc, Leclerc Papers, 239 K 4, Private Archives, reproduced in its entirety in Bodinier, *Le retour de la France en Indochine*, 175–83.

37 8 February 1946, between Salan and Ho Chi Minh, Leclerc Papers, 239 K 5, Private Archives, reproduced in its entirety in Bodinier, *Le retour de la France en Indochine*, 202–4.

38 12 December 1946, report from Morlière on the situation in north Indochina, Leclerc Papers, 239 K 5, Private Archives, reproduced in its entirety in Bodinier, *Le retour de la France en Indochine*, 365–72.

39 7 September 1945, memo from Leclerc to de Guillebon, National Defence Staff Files, Series 4 Q (111), Military Archives.

40 15–22 March 1945, situation in Indochina, Files AG3(4), National Archives.

41 18 July 1945, situation in Indochina on 14 July, Files AG3(4), National Archives.

42 13 March 1947, *Journal Officiel de la République Française*, No. 24, Session No. 28, National Assembly.

43 11 November 1945, note to Pompidou on situation in Indochina, Files AG3(4), National Archives.

44 5 December 1947, note from Morlière on Indochina, Files F/60/3035, Indochina Committee Folder, "brides de données de séances," National Archives.

45 Wainstock and Miller, *Indochine et Vietnam.*

46 Note from a navy officer on situation in Indochina (January–September 1947), Files F/60/3035, Indochina Committee Folder, "brides de données de séance," National Archives.

47 7 November 1945, note to Pompidou, Files AG3(4), National Archives.

48 25 May 1946, report from Juin on his mission in the Far East, Leclerc Papers, 238 K 5, Private Archives, reproduced in its entirety in Bodinier, *Le retour de la France en Indochine*, 261–6.

49 10 December 1946, report from Debès on events of 20 November 1946, Indochina Files, Series 10 H (602-2), Military Archives.

50 15 July 1947, letter from Valluy on Ho Chi Minh's operational activity, Indochina Files, Series 10 H (950), Military Archives.

51 2 January 1947, *Le Monde.*

52 10–16 November 1947, a summary of military events from Major-General Jean Humbert, Files F/60/3035, National Archives.

53 5 December 1947, note on Indochina, Files F/60/3035, Indochina Committee Folder, 'brides de données de séance,' National Archives.

54 1 February 1947, memo from Valluy for reinforcement soldiers, Indochina Files, Series 10 H (297), Military Archives.

55 29 November 1946, intervention by Bidault, Files F/60/3035, Indochina Committee Folder, "brides de données de séances," National Archives.

56 18 March 1947, *Journal Officiel de la République Française*, No. 26, Session No. 30, National Assembly.

57 18 March 1947, *Journal Officiel de la République Française*, No. 26, Session No. 30, National Assembly.

58 Lakoff, *The Political Mind: Why You Can't Understand 21st-Century*; *Moral Politics*; *The Political Mind: A Cognitive Scientists' Guide*.

59 20 October 1945, note for fighters in the 2nd division on their conduct, Indochina Files, Series 10 H (569), Military Archives.

60 25 May 1946, report from Juin on the mission in the Far East, Leclerc Papers, 238 K 6, Private Archives, reproduced in its entirety in Bodinier, *Le retour de la France en Indochine*, 261–6.

61 22 October 1946, note from Pignon, Indochina Files, Series 10 H (143), Military Archives.

62 3 February 1947, note from Valluy about repercussions of events in Hanoi, Indochina Files, No.17, Series 10 H (297), Military Archives.

63 10 October 1945, summary of report from Leclerc, Files AG3(4), National Archives.

64 7 November 1945, note for Pompidou about situation in Indochina, Files AG3(4), National Archives.

65 Dalloz, *La guerre d'Indochine*; Valette, *La Guerre d'Indochine*; Franchini, *Les Guerres d'Indochine*.

66 28 August 1945, report from Leclerc to d'Argenlieu, National Defence Staff Files, Series 4 Q (78), Military Archives.

67 19 October 1946, Indochina Files, Series 10 H (165), Military Archives.

68 10 December 1946, report on events of 20 November 1946, Indochina Files, Series 10 H (602-2), Military Archives.

69 See Bulletin Study 48, about situation leading up to operations starting on 20 November, Files F/60/3035, Indochina Committee Folder, "brides de données de séances," National Archives.

70 10 December 1946, report on events of 20 November 1946, Indochina Files, Series 10 H (602-2), Military Archives.

71 11 December 1947, Indochina Files, Series 10 H (3534), Military Archives.

72 25 May 1946, report from Juin, Lerclerc Papers, Private Archives, reproduced in its entirety in Bodinier, *Le retour de la France en Indochine*, 261–6.

73 6 January 1947, report from Leclerc on his mission in the Far-East, Files F/60/3035, Inter-ministerial Conferences Folder, National Archives.

74 24 November 1946, telegram from d'Argenlieu to Valluy, Indochina Files, Series 10 H (98), Military Archives.

75 8 September 1947, note on the attitude of Viet Minh leaders regarding possible negotiations, Indochina Files, Series 10 H (625), Military Archives; see also note from Valluy, "répercussions des évènements de Hanoi sur la ligne de conduite française dans les pays annamites," Indochina Files, No. 17, Series 10 H (297), Military Archives.

76 Valette, *La Guerre d'Indochine*, 75–6.

77 18 March 1947, *Journal Officiel de la République Française*, No. 26, Session No. 30, National Assembly.

78 8 December 1946, Bulletin Study 48, "brides de données de séances"; "situation ayant conduit aux opérations commencées le 20 novembre 1946 au Tonkin," Files F/60/3035, Indochina Committee Folder, "brides de données de séances," National Archives.

79 8 February 1946, Leclerc Papers, 239 K 5, Private Archives, reproduced in its entirety in Bodinier, *Le retour de la France en Indochine*, 202–4.

80 5 December 1946, National Defence Staff Files, Series 4 Q (1), National Defence Historical Service.

81 Dalloz, *La guerre d'Indochine*; Ruscio, *La Guerre Française d'Indochine*; Wainstock and Miller, *Indochine et Vietnam*.

82 13 March 1946, from the government to the high commissioner, Indochina Committee Folder, National Defence Staff Files, Series 4 Q (77), Military Archives.

83 11 September 1945, Briefing Note 13, Files AG3(4), National Archives.

84 26 April 1946, memo from the high commissioner about the turning political point in Indochina, Files F/60/3035, Indochina Committee Folder, "brides de données de séance," National Archives.

85 11 March 1947, *Journal Officiel de la République Française*, No. 23, Session No. 27, National Assembly.

86 18 March 1947, *Journal Officiel de la République Française*, No. 26, Session No. 30, National Assembly.

87 12 December 1946, *Journal Officiel de la République Française*, No. 18, Session No. 7, National Assembly.

88 6 May 1947, *Journal Officiel de la République Française*, written questions, No. 38, Session No. 49, National Assembly.

89 22 March 1947, *Journal Officiel de la République Française*, No. 29, Session No. 34, National Assembly.

90 Opposition over the understanding and definition of virility was captured in military documents, but it could be found among other actors given the general political context and the overall tendency of the dataset.

91 1 January 1947, Indochina Files, Series 10 H (161), Military Archives.

92 5 February 1946, Files F/60/3035, Indochina Committee Folder, "brides de données de séances," National Archives.

93 Franchini, *Les Guerres d'Indochine*; Wainstock and Miller, *Indochine et Vietnam*.

94 November 1946, in Dalloz, *La guerre d'Indochine*, 112.
95 Ruscio, *La Guerre Française d'Indochine.*
96 5 December 1947, Files F/60/3035, Indochina Committee Folder, "brides de données de séances," National Archives.
97 12 December 1946, report from Morlière about situation in northern Indochina, 239 K 5, reproduced in its entirety in Bodinier, *Le retour de la France en Indochine*, 365–72.
98 Franchini, *Les Guerres d'Indochine.*
99 15 September 1945, political instructions from d'Argenlieu, Indochina Files, Series 10 H (161), Military Archives.
100 18 July 1945, situation in Indochina as at 14 July, Files AG3(4), National Archives.
101 15 May 1947, Files F60/3035, Indochina Committee Folder, National Archives.
102 20 May 1947, Indochina Files, Series 10 H (3828), Military Archives.

Chapter Five

1 Hoffman and Hoffman, *Sovereignty in Dispute*; Middlebrook, *Operation Corporate.*
2 Middlebrook, *Operation Corporate.* There is no actual evidence that the junta was behind Davidoff's moves, but it took advantage of the situation nonetheless.
3 Oakes, "Diversionary War and Argentina's Invasion of the Falklands Islands"; Dufour, *Un siècle de crises internationales.*
4 Train, "An Analysis of the Falklands/Malvinas Islands Campaign."
5 Hoffman and Hoffman, *Sovereignty in Dispute*; Vacs, "A Delicate Balance: Confrontation and Cooperation between Argentina and the United States."
6 Freedman and Gamba, *Signals of War.*
7 12 March 2013, Falkland Islands Government, Results of the referendum on the Political Status of the Falkland Islands. [Press release], 12 March 2013. Available at http://www.falklands.gov.fk/results-of-the-referendum-on-the-political-status-of-the-falkland-islands/.
8 Marx, *Histoire de l'Angleterre.*
9 These include works from decision-makers and military personnel involved in the conflict (Woodward, *One Hundred Days*; Bramley, *Excursion to Hell*; Thatcher, *The Downing Street Years*) as well as works from historians seeking to explain the succession of events that led to the war, how military operations were carried out (Hoffman and Hoffman, *Sovereignty in Dispute*; Freedman, "The War of the Falkland Islands"; *Britain and the Falklands War*; Middlebrook, *Operation Corporate*; Kinney, *National Interest/National Honor*; Hasting and Jenkins, *The Battle for the Falklands*), and what the legal ramifications of the conflict were in relation to international law (Calvert, "Sovereignty and the Falklands Crisis"; Franck, "Dulce et decorum est"; Perl and Larson, *The Falklands Dispute in International Law*; Windsor, "Diplomatic Dimension of the Falklands Crisis"; Schmitt and Green,

"The Falklands Crisis and the Laws of War"; Bluth, "The British resort to force in the Falklands/Malvinas Conflict").

10 Rotberg and Rabb, *The Origins and Prevention of Major Wars*; Paul, *Asymmetric Conflicts.*

11 Gibran, *The Falklands War*; Welch, *Justice and the Genesis of War*; Welch, *Painful Choices.*

12 Levy and Vakili, "Diversionary Action by Authoritarian Regime"; Welch, "Remember the Falklands?"; *Painful Choices*; Sobek, "Rallying around the Podesta"; Weisiger, *Logics of War.*

13 Norpoth, "The Falklands War and Government Popularity"; Sanders, Ward, Marsh, and Fletcher, "Government Popularity and the Falklands War"; Gustafson, *The Sovereignty Dispute over the Falkland (Malvinas) Islands*; Lai and Reiter, "Rally 'round the Union Jack?"; Moore, *Margaret Thatcher*; Sharp, *Thatcher's Diplomacy*; Schenoni, Braniff, and Battaglino, "Was the Malvinas/Falklands a Diversionary War?"

14 Lebow, "Miscalculation in the South-Atlantic"; Gibran, *The Falklands War*; McCourt, "Role-Playing and Identity Affirmation"; *Britain and World Power*; Grandpierron, "Preserving Great Power Status."

15 Sanders, *Losing an Empire*; Marsh, "Anglo-American Relations 1950–1951."

16 Marx, *Histoire de l'Angleterre*; Chassaigne, *La Grande Bretagne et le monde.*

17 Robbins, *The Eclipse of a Great Power.*

18 McCourt, "Role-Playing and Identity Affirmation," 1618.

19 Moore, *Margaret Thatcher,* 656.

20 Ibid., 537.

21 Ibid., 533.

22 Ibid., 529.

23 Dufour, *Un siècle de crises internationales.*

24 Peterson, "Soviet Military Objectives in the Arctic Theater."

25 Thatcher, *The Downing Street Years*, 173–4.

26 Gustafson, *The Sovereignty Dispute.*

27 Moore, *Margaret Thatcher.*

28 Ibid., 676.

29 25 November 1979, the "Carless" letter to Harding, Thatcher Foundation.

30 16 November 1980, "The Falklands: Argentine Press Coverage." Thatcher Foundation.

31 Ellerby, "The Role of the Falkland Lobby."

32 Dodds, *Pink Ice.*

33 Donaghy, *The British Government and the Falklands Islands.*

34 John Ure is a British diplomat, Ambassador to Cuba (1979–81) before becoming Assistant under Secretary of State at the Foreign and Commonwealth Office.

35 10 May 1979, "Falkland Islands," minute from Ure to Ridley. Thatcher Foundation. Nicholas Ridley was minister of state at the Foreign and Commonwealth Office. He

was specifically in charge of policies concerning the Falkland Islands. After visiting the islands in 1979, he concluded that Britain could not afford to defend them.

36 25 January 1980, "Falkland Islands," OD (79)31. Thatcher Foundation.

37 Gustafson, *The Sovereignty Dispute.*

38 Gibran, *The Falklands War.*

39 *Phase 1*: 3.4 per cent; *Phases 2* and *3*: 0 per cent.

40 2 September 1979, minute from Hunt to Thatcher, Thatcher Foundation.

41 21 May 1980, minute from Ridley to Carrington, Thatcher Foundation.

42 14 September 1981, minute from Leon Britton to Thatcher, Thatcher Foundation.

43 10 May 1979, minute from Ure to Ridley, Thatcher Foundation.

44 24 September 1979, minute from Chancellor of Exchequer Denis Healey to Thatcher, Thatcher Foundation. See also 8 April 1982, "Gibraltar Effect on the Falklands Crisis," letter from FCO to 10 Downing Street, Thatcher Foundation.

45 20 January 1980, FCO Minister of State, OD Committee, Thatcher Foundation.

46 Taliaferro, "Power Politics and the Balance of Risk."

47 8 April 1982, letter from No. 10 to FCO about Defence and Overseas subcommittee meeting of 7 April, Thatcher Foundation.

48 Hoffman and Hoffman, *Sovereignty in Dispute*; Moore, *Margaret Thatcher.*

49 Middlebrook, *The Falklands War*; Moore, *Margaret Thatcher*. The use of the Island was absolutely vital for the British. It has a very long runway from which strategic bombers could operate, was the last stop before reaching the Falklands.

50 3 April 1982, minutes of Ministry of Defence Chiefs of Staff committee, COS 3rd meeting 1982, Thatcher Foundation.

51 14 April 1982, No. 10 record of conversation between Thatcher and Haig, full transcript of 1st call, Thatcher Foundation.

52 13 May 1983, White House record, Thatcher Foundation.

53 8 April 1982, No. 10 record of dinner conversation between Thatcher and Haig, Thatcher Foundation.

54 The code was also found in internal documents, where it was associated to a different idea: the risk that if Argentina got closer to the USSR, it might receive missiles and aircraft from the Soviets.

55 Moore, *Margaret Thatcher.*

56 Hastings and Jenkins, *The Battle for the Falklands*; Hoffman and Hoffman, *Sovereignty in Dispute*; Moore, *Margaret Thatcher.*

57 8 April 1982, No. 10 record of dinner conversation between Thatcher and Haig, Thatcher Foundation.

58 14 April 1982, record of No. 10 telephone conversation between Thatcher and Haig, full transcript of 2nd call, Thatcher Foundation.

59 29 April 1982, letter from Thatcher to Reagan, Thatcher Foundation.

60 19 April 1982, OD and Defence Subcommittee, Thatcher Foundation.

61 4 May 1982, letter from Thatcher to Reagan, Thatcher Foundation.

62 This section is based and expands on previous research published by the *Round Table* (Grandpierron, "The British Use of Domestic Political Situations").
63 Sharp, *Thatcher's Diplomacy*, 65.
64 It is important to remember that Thatcher was an outsider when she was elected leader of the party in 1979 (Buckley, *The Prime Minister and Cabinet*).
65 Gustafson, *The Sovereignty Dispute*; Sharp, *Thatcher's Diplomacy.*
66 15 October 1979, note from No. 10 to FCO, Thatcher Foundation.
67 30 March 1982, minute from FCO to Thatcher, Thatcher Foundation.
68 30 March 1982, minutes of Defence Operations Executive meeting, Thatcher Foundation.
69 30 March 1982, minute from FCO to Thatcher, Thatcher Foundation.
70 30 March 1982, minutes of Defence Operations Executive meeting, Thatcher Foundation.
71 30 March 1982, minute from FCO to Thatcher, Thatcher Foundation.
72 30 March 1982, minutes of Defence Operations Executive meeting, Thatcher Foundation.
73 6 April 1982, Prime Minister's Questions, Hansard Digest, HC 2/822-26, Column 822.
74 Landry, "L'analyse de contenu."
75 2 April 1982, 09:45, Cabinet minutes, Thatcher Foundation; see also 2 April 1982, 19:30 cabinet minutes, Thatcher Foundation.
76 3 April 1982, notes from Gow of 1922 Committee Meeting, Thatcher Foundation.
77 20 September 1979, minute from Carrington to Thatcher, Thatcher Foundation.
78 3 December 1980, OD(80) 25th, Thatcher Foundation.
79 Gibran, *The Falklands War.*
80 Peaking at 18.6 per cent in *Phase 1* and increasing to 19 per cent in ministerial documents alone.
81 8 April 1982, (both quotes) No. 10 record of dinner conversation between Thatcher and Haig, Thatcher Foundation.
82 Moore, *Margaret Thatcher.*
83 12 April 1982, No. 10 record of conversation between Thatcher and Haig, 3rd meeting, Thatcher Foundation.
84 14 April 1982, record of 2nd No. 10 telephone conversation between Thatcher and Haig, Thatcher Foundation.

Chapter Six

1 Jervis, *The Logics of Images.*
2 Evergetism, or doing good deeds, is a social practice captured by the Latin expression *panem et circences* ("bread and games"), in which wealthy Romans, then victorious generals and later emperors, offered circus games to flatter the crowd's ego

(these particularly gladiatorial battles proved very efficient in satisfying the Roman passion for violence and blood; see Vayne, "Panem et circenses"; Zuiderhoek, *The Politics of Munificence*). A possible modern version of *panem et circenses* would link a leader's political fate to status and virility. Depending on how the ideas of strength, bravery, and national status are understood within public opinion, leaders wanting to remain in power would be implicitly forced to choose to go to war.

3 Identified in 0.5 per cent of documents.

4 15 April 1982, letter to Conservative Constituency chairman, MPs and others, Thatcher Foundation.

5 4 May 1982, full cabinet minutes, conclusions of meeting, CC(82)22nd, Thatcher Foundation.

6 On this matter, good examples can be found: 10 May 1979, minute from Ure to Ridley; 19 January 1982, memo from Hunt to Carrington, Thatcher Foundation.

7 7 January 1982, letter from Thatcher to Sir Richard Needham, Thatcher Foundation.

8 6 April 1982, letter from Thatcher to Reagan, Thatcher Foundation.

9 4.3 per cent in Thatcher's documents compared to 2.2 per cent and 2.6 per cent for ministers and public respectively.

10 *Phase 1*: 26.2 per cent; *Phase 2*: 20.9 per cent; *Phase 3*: 36.2 per cent.

11 28 April 1980, Anglo-Argentine talks on the Falkland Islands, Thatcher Foundation.

12 1 July 1980, Armstrong briefing to Thatcher, for the first quote; 29 January 1980, remarks from FCO Secretary during Defence and Overseas Committee, for the second quote, Thatcher Foundation.

13 5 April 1982, FCO report of conversation between Carrington and Hunt, Thatcher Foundation (for both quotes).

14 8 April 1982, No. 10 record of Thatcher-Haig talks, Thatcher Foundation.

15 Harriers and Sea Harriers (VTOL aircrafts designed to provide ground support) were also tasked with providing air defence for the fleet and engaging the Argentine air force, missions for which they were not designed.

16 Middlebrook, *The Falklands War*.

17 27 May 1982, Secretary of State for Defence, full cabinet minutes, CC(82)30th, Thatcher Foundation.

18 15 April 1982, letter from Parkinson to Conservative Constituency chairman, MPs and others, Thatcher Foundation.

19 8 April 1982, No. 10 record of conversation between Thatcher and Haig, Thatcher Foundation.

20 Gibran, *The Falklands War*.

21 25 May 1982, Pym minute to MT, Thatcher Foundation.

22 10 May 1981, Fearn letter to UK Ambassador in Buenos Aires, Thatcher Foundation.

23 23 February 1981, FCO record of Anglo-Argentine talks on the Falkland Islands.

24 3 December 1980, Defence and Overseas Committee, OD(80)25th, Thatcher Foundation.
25 14 April 1982, all quotes from confidential annexe of full cabinet minutes, CC(82) 17th, Thatcher Foundation.
26 21 April 1982, No. 10 record of conversation between Thatcher and Ahidji, Thatcher Foundation.
27 16 April 1982, message from Margaret Thatcher to Ronald Reagan, Thatcher Foundation.
28 5 May 1982, letter to Fearn from British Ambassador in Buenos Aires, Thatcher Foundation.
29 26 June 1981, Port Stanley to FCO, Thatcher Foundation.
30 21 January 1982, draft response to parliamentary question from Michael Shersby MP, Thatcher Foundation.
31 Middlebrook, *The Falklands War*.
32 31 March 1982, FCO to British Embassy in Buenos Aires, Thatcher Foundation.
33 22 March 1982, Hunt to FCO, Baron Buxton of Alsa's talks with Mendez, Thatcher Foundation.
34 23 March 1982, note from John Coles to Thatcher, Thatcher Foundation.
35 6 April 1982, full cabinet minutes, CC(82)16th, Thatcher Foundation.
36 20 May 1982, full cabinet minutes, CC(82)28th, Thatcher Foundation.
37 1 July 1980, Defence and Overseas Committee, "Falkland Islands," OD(80)46, Thatcher Foundation.
38 See also MP Winston Churchill (grandson of) intervention, 7 April 1982, House of Commons Digest, Vol. 21, cc.959–1052, Thatcher Foundation.
39 3 July 1982, speech from Thatcher to Conservative rally at Cheltenham, https://www.margaretthatcher.org/document/104989.
40 Taliaferro, "Power Politics and the Balance of Risk."
41 29 November 1980, debate with councillors over Ridley's visit to Falklands, Thatcher Foundation.
42 Parr, "Representations of Grief and the Falklands War."
43 27 March 1982, message from UK Embassy in Buenos Aires to FCO, Thatcher Foundation.
44 30 March 1982, message from FCO to UK Embassy in Washington DC, Thatcher Foundation.
45 The 1922 Committee meeting is also known as the Conservative Private Members' Committee, the Conservative Parliamentary Group at the House of Commons.
46 3 April 1982, all questions asked can be found in Gow's notes of the 1922 Committee Meeting, Thatcher Foundation.
47 2 April 1982, *HMS Endurance* to CINCFLEET [Commander-in-Chief of Fleet], Thatcher Foundation.

48 2 April 1982, FCO to UK Embassies and Missions, Thatcher Foundation.

49 Middlebrook, *Operation Corporate*, 52. This event clearly demonstrates Ty Solomon's argument that "emotions and the body are not two separate analytical issues, but are rather bound together through the symbolic politics of discourse" (Solomon "Embodiment, Emotions and Materialism," 58). See also Sontag, *On Photography*; *Regarding the Pain of Others*; Bleiker and Kay, "Representing HIV/AIDS in Africa"; and Gregory, "Photographing War."

50 5 May 1981, letter from UK Ambassador in Buenos Aires to Fearn, Thatcher Foundation.

51 10 May 1979, minute from Ure to Ridley, Thatcher Foundation.

52 31 May 1982, message from FCO to UK Embassy in Buenos Aires, Thatcher Foundation.

53 7 April 1982, message from UK Embassy in Washington to FCO, Thatcher Foundation.

54 5 April 1982, letter from FCO to No. 10, Thatcher Foundation.

55 14 June 1982, transcript of TV interview, BBC Radio News, Thatcher Foundation.

56 Bellot, "The Malvinas/Falklands War."

57 Dyson, *The Blair Identity*, 40.

58 7 January 1981, letter from Thatcher to Needham, Thatcher Foundation.

59 See for example: 19 August 1981, letter from Thatcher to Buxton, Thatcher Foundation.

60 2 April 1982, 09.45 a.m., cabinet meeting, 09.45 a.m., Thatcher Foundation.

61 14 April 1982, Thatcher speech in the House of Commons, Hansard, HC 21/1146-50, Thatcher Foundation.

62 3 April 1982, Prime Minister's Questions, House of Commons, Hansard HC 21/633–38.

63 10 June 1982, Thatcher ITN interview, Thatcher Foundation.

64 7 April 1982, No. 10 record of conversation between Thatcher and Schmidt, Thatcher Foundation.

65 7 April 1982, No. 10 record of conversation between Thatcher and Schmidt, Thatcher Foundation.

66 11 May 1982, No. 10 record of conversation between Thatcher and Reagan, Thatcher Foundation.

67 25 May 1982, message from Thatcher to President Turbay of Colombia, Thatcher Foundation.

68 Dyson, *The Blair Identity*.

69 2 April 1982, Thatcher ITV interview, Thatcher Foundation.

70 13 April 1982, No. 10 record of dinner conversation between Thatcher and Haig, Thatcher Foundation.

71 5 April 1982, Thatcher ITN interview, Thatcher Foundation.

72 9 April 1982, FCO draft reply from Thatcher to Chairman of British Community Council in Buenos Aires, Thatcher Foundation.

73 5 April 1982, Thatcher ITN interview, Thatcher Foundation.

74 Some have argued that Thatcher ordered the sinking of the Argentine ship, *General Belgrano*, even though it was, at the time the decision was taken, outside the maritime exclusion zone, in order to force Argentina to reject the Peruvian peace proposal. I found no evidence of this in the declassified documents, although I did find documents advising to authorize the extension of Royal Navy rules of engagement based on the evolution of Argentina's naval deployments and in the eventuality that Argentina officially declared war on the UK, which would have legal implications. See: 2 May 1982, No. 10 record of conversation between Thatcher and William Whitelaw, John Nott, Cecil Parkinson, Michael Havers, Sir Terence Lewin, Admiral John Fieldhouse, and Sir Antony Acland, Thatcher Foundation; and 2 May 1982, minutes of Defence and Overseas subcommittee on South Atlantic and the Falklands (SA) (82)23rd, Thatcher Foundation.

75 Both quotes from 22 April 1982, confidential annex of full cabinet minutes, Thatcher Foundation.

76 2 June 1982, Thatcher ITV interview, Thatcher Foundation.

77 17 May 1982, No. 10 record of conversation between Thatcher and François Mitterrand, Thatcher Foundation.

78 27 May 1982, Thatcher message to President Lopez Portello of Mexico.

79 13 May 1982, White House record of conversation between Reagan and Thatcher, Thatcher Foundation; see also 19 May 1982, No. 10 record of telephone conversation between Thatcher and UN Secretary General Javier Pérez de Cuéllar and 27 May 1982, Thatcher message to President Portillo of Mexico, Thatcher Foundation.

80 14 April 1982, No. 10 record of conversation between Thatcher and Haig, full transcript of 2nd call, Thatcher Foundation.

81 11 June 1982, letter from Thatcher to Michael Foot, Thatcher Foundation.

82 2 June 1982, Thatcher ITV interview, Thatcher Foundation.

83 14 September 1981, minute from Britton to Thatcher, Thatcher Foundation.

84 18 September 1979, "redrafting minute to Thatcher on options for the Falklands, need to frame leaseback positively," minute from Lyne to Duggan, Thatcher Foundation.

85 15 February 1982, response from Thatcher to Carrington, Thatcher Foundation.

86 15 February 1982, a representative example in a letter from No. 10 to FCO, Thatcher Foundation.

87 21 September 1979, "Falkland Islands," a representative example in minute from Hunt to Thatcher, Thatcher Foundation.

88 12 October 1979, letter from FCO to Thatcher, Thatcher Foundation.

89 Ibid.

90 12 October 1979, letter from FCO to Thatcher, Thatcher Foundation.

91 16 April 1982, "I have never pretended to agree with you about everything," letter from Carrington to Thatcher, Thatcher Foundation. Thatcher was not keen to accept his resignation. The argument put forward by Carrington for his resignation was his mismanagement of the negotiations with Argentina and the contradictory signals he had sent about British resolve. The general tone of the letter and the way he ends it suggest that any opposition was not personal: "My admiration for your courage and determination and resourcefulness is unbounded."

92 Armstrong, Cabinet Secretary's notebook, Cabinet Office Records, Cabinet Secretary's Office, National Archives.

93 25 May 1982, letter from Thatcher to Turbay, Thatcher Foundation.

94 5 May 1982, note from Armstrong to cabinet, CC(82) 21, Thatcher Foundation.

95 4 May 1982, letter from Thatcher to Reagan, Thatcher Foundation.

96 7 May 1982, letter from Thatcher to Chief Rabbi Jakobovits, Thatcher Foundation.

97 5 April 1982, FCO record of conversation between Carrington and Hunt, Thatcher Foundation.

98 8 April 1982, No. 10 record of conversation between Thatcher and Haig, Thatcher Foundation.

99 Middlebrook, *The Falklands War*.

100 29 April 1982, letter from Thatcher to Denis Skinner MP, Thatcher Foundation.

101 24 May 1982, letter from Thatcher to Brigadier Harbottle, Thatcher Foundation.

102 25 May 1982, both quotes in letter from Thatcher to Turbay, Thatcher Foundation.

103 20 May 1982, full cabinet minutes, CC(82)28th, Thatcher Foundation.

104 13 April 1982, No. 10 record of conversation between Thatcher and Haig, Thatcher Foundation.

105 29 April 1982, letter from Thatcher to Reagan, Thatcher Foundation.

106 De Larrinaga and Doucet, "Governmentality, Sovereign Power and Intervention."

107 10 June 1982, Thatcher ITN interview, Thatcher Foundation.

108 10 June 1982, Thatcher BBC interview, Thatcher Foundation.

109 7 April 1982, No. 10 record of telephone conversation between Thatcher and Schmidt, Thatcher Foundation.

110 5 April 1982, Thatcher ITN interview, Thatcher Foundation.

111 Breuning, *Foreign Policy Analysis*.

112 Sharp, *Thatcher's Diplomacy*.

113 Hastings and Jenkins, *The Battle for the Falklands*, 317.

114 Ibid.; Middlebrook, *Operation Corporate*.

115 Hastings and Jenkins, *The Battle for the Falklands*, 317.

116 Gibran, *The Falklands War*, 110.

Chapter Seven

1 Gilmore, *The Grenada Intervention*; O'Shaughnessey, *Grenada: An Eyewitness Account;* Crandall, *Gunboat Diplomacy.*

2 Payne, Sutton, and Thorndike, *Grenada Revolution and Invasion*; Cornell-Smith, "The Grenada Invasion in Historic Perspective"; O'Shaughnessey, *Grenada: An Eyewitness Account*; Gilmore, *The Grenada Intervention*; Burrowes, *Revolution and Rescue Grenada*; Williams, "Grenada Five Years Later"; Adkin, *Urgent Fury the Battle for Grenada*; Haass, *Intervention*; Crandall, *Gunboat Diplomacy.*

3 Anno and Einspach, "Command and Control and Communications Lessons Learned"; Cole, *Operation Urgent Fury Grenada*; Lewis, Williams, and Clegg, *Grenada: Revolution and Invasion*; Clarke, *Operation Urgent Fury.*

4 Rubner, "The Reagan Administration, the 1973 War Powers Resolution and the Invasion of Grenada."

5 Joyner, "Reflections on the Lawfulness of Invasion"; Gilmore, *The Grenada Intervention*; Wheeler, "The Grenada Invasion."

6 Crandall, *Gunboat Diplomacy.*

7 Haass, *Intervention*; Schweizer, *Reagan's War*; Crandall, *Gunboat Diplomacy.*

8 O'Shaughnessey, *Grenada: An Eyewitness Account*; Burrowes, *Revolution and Rescue Grenada.*

9 Adkin, *Urgent Fury the Battle for Grenada.*

10 Weber, *Simulating Sovereignty*; *Faking It.*

11 Yahri-Milo, *Who Fights for Reputation.*

12 Go, "American Decline and Performative War or How to Do Things Done."

13 Schweizer, *Reagan's War*; Gueniffrey and Lentz, *La fin des empires.*

14 Burrowes, *Revolution and Rescue Grenada*, 73.

15 Haass, *Intervention*; Welch, *US-Caribbean Relations.*

16 Kagan, *A Twilight Struggle.*

17 The New Jewel Movement or New Joint Endeavor to Welfare, Education and Liberation, is a Marxist-Leninist political party founded in 1973 by Maurice Bishop who took power in 1979 after a coup. The party then suffered from an internal division resulting in a radical section seizing power, deposing Bishop, and executing him.

18 Gilmore, *The Grenada Intervention.*

19 Crandall, *Gunboat Diplomacy*, 135.

20 Seabury and McDougall, *The Grenada Papers.*

21 Schweizer, *Reagan's War*; Crandall, *Gunboat Diplomacy.*

22 Gilmore, *The Grenada Intervention.*

23 The Caribbean Basin Initiative, an economic program conceived in 1982 and validated by congress in 1983 (to be put in practice in 1984), was presented as an

economic aid program in favour of the Caribbean region. In practice, the program also aimed at strengthening the US control over the region.

24 Payne, Sutton, and Thorndike, *Grenada Revolution and Invasion.*

25 Reagan, *An American Life*, 314; Reagan, *The Reagan Diaries*, edited by Douglas Brinkley, 87–8.

26 Burrowes, *Revolution and Rescue Grenada*, 140.

27 Undated, "The Strategic and Economic Importance of the Caribbean Sea Lanes," Douglas Holladay Files, Office of Public Liaison, Series I, General Subject File OAS 12/51, 12/52, Ronald Reagan Library; see also 16 December 1983, "The Strategic and Economic Importance of the Caribbean Sea Lanes," Crisis Management Center (CMC), NSC Records 1981–85, RAC Box 8, Reagan Library.

28 12 October 1983, "Grenada Invasion Assessment," Holladay Files, Office of Public Liaison, Series I, General Subject File OAS 12/51, 12/52, Reagan Library.

29 Undated, "Grenada - Urgent Fury (03/03/1983–10/21/1983)," Under Secretary Eagleburger to Jonathan Howe, Folder, Box 90500, ID #37034, Jacqueline Tillman Files, Reagan Library.

30 Undated, "Soviet objectives and intentions in Latin America," General Subject File OAS 12/51, Office of Public Liaison, Series 1, Holladay Files, Reagan Library.

31 Payne, Sutton, and Thorndike, *Grenada Revolution*; Crandall, *Gunboat Diplomacy*; 21 April 1983, "Soviet Objectives and Intentions in Latin America," CIA intelligence, Series 1, Subject File RAC Box 8, Roger Fontaine Files, Reagan Library; see also Office of Public Liaison, Series 1, General Subject File OAS 12151, Holladay Files, Reagan Library.

32 As a reminder, *Phase 1* goes from 1 January 1980 (the start of Reagan's presidency) to 18 October 1983 (the day before the coup in Grenada that overthrew Bishop). *Phase 2* goes from 19 October to 24 October 1983 (the day of the coup in Grenada to the day before US troops landed). *Phase 3* goes from 25 October to 31 December 1983 (the end of the war).

33 21 April 1983, report, Directorate of Intelligence, CIA, Box 5, Alfonso Sapia-Bosch Files, Reagan Library.

34 25 October 1983, memo, Country File, Latin America, Grenada, Box 3, Folder 1 (6/17/83–10/27/83), Executive Secretariat Files, Reagan Library.

35 Cable to secretary of state, NSC Cable File Records 1982–85, Box 42, Executive Secretariat Files, Reagan Library.

36 25 October 1983, NSC Cable File Records 1982–85, Box 42, Executive Secretariat Files, Reagan Library.

37 Payne, Sutton, and Thorndike, *Grenada Revolution.*

38 Schweizer, *Reagan's War*, 208.

39 Crandall, *Gunboat Diplomacy.*

40 "Necessary Invasion," OA 9450, Edwin Meese files, Reagan Library.

41 21 September 1982, subject file, Series I, RAC Box 8, Fontaine Files, Reagan Library.

42 21 April 1983, Box 5, Sapia-Bosch Files, Reagan Library.

43 Undated, 1981, "Grenada – Urgent Fury (03/03/1983–10/21/1983)," Grenada Report Folder, Box 90500, Tillman Files, Reagan Library.

44 "Review Grenada," Subject File, Series 1, RAC Box 8, Fontaine Files, Reagan Library.

45 Undated, 1981, "Review Grenada," Subject File, Series 1, RAC Box 8, Fontaine Files, Reagan Library.

46 26 October 1983 "An Easy but Costly Invasion," *Hartford Courant*, OA 9450, Meese Files, Reagan Library.

47 1 June 1983, WHORM: CO Countries Files, CO 058 Greece 280000–CO 058 Grenada 201499, Box 80, Reagan Library.

48 Payne et al., *Grenada Revolution and Invasion.*

49 18 September 1981, letter from Max Friedensdorf (assistant to the President) to Micky Leland, WHORM: CO Countries Files, CO 058 Greece 280000–CO 058 Grenada 201499, Box 80, Reagan Library.

50 26 October 1983, *New York Times*, OA 9450, Meese files, Reagan Library.

51 25 October 1983, White House summary, update on wires summary, OA 9450, Meese Files, Reagan Library.

52 Undated, "Students Held Hostage," press article (newspaper not identified), WHO Records, OA 10495, 10446, 10047, Public Affairs Office Files, Reagan Library.

53 24 October 1983, *CBS* and the *New York Times* poll regarding the question, *Were the Americans in Grenada in danger before the troops were sent or not?*, OA 11244, Box 10, Michael Barooday Files, Reagan Library.

54 27 October 1983, *Baltimore Evening Sun*, White House news summary, Series VI: mail samples 04/10/1983–06/09/1983, Box 75, Anne Higgins Files, Reagan Library.

55 Undated, "Has Anybody Seen Ronald Reagan?," Patrick Buchanan, FG GOT-ORG FG001, President of the US Files, 143717(2 of 3)–158999, Box 48.

56 26 October 1983, possibly Richard Utter (document very hard to read), *Philadelphia Daily News*, OA 9450, Meese Files, Reagan Library.

57 28 October 1983, RAC Box 6, Chron. file (official) 10/28/1983–11/07/1983, McFarlane Files, Reagan Library.

58 Undated, NSC Subject File, Box 52, Executive Secretariat Files, Reagan Library.

59 10 November 1983, memo from CIA to President, NSC File Records, RAC Box 7, Executive Secretariat Files, Reagan Library.

60 31 October 1983, 12.30 p.m., "Briefing for Freedom Federation and Ethnic Leaders," "Grenada – Urgent Fury (03/03/1983–10/21/1983)," Box 90499–90500, File Folder, Tillman Files, Reagan Library.

61 CO 058 Greece 280000–CO 058 Grenada 201499, Box 80, CO Country Files, Reagan Library.

Chapter Eight

1 Thucydides, *History of the Peloponnesian War*, Book 5, chapters 84–116.

2 12 February 1983, memo from McFarlane to Clark, CHRON Files, RAC Box 5, 123662, McFarlane Files, Reagan Library.

3 2 January 1983, memo from Craig Fuller to Dave Gergen, FG GOT-ORG, FG001, 111672-119734, President of the US Files, Reagan Library.

4 27 October 1983, letter from rescued student to Reagan, Series VI: mail samples 04/10/1983–06/09/1983, Box 75, Higgins Files, Reagan Library.

5 26 October 1983, "Reagan Invades Grenada in a Perilous Move," *Philadelphia Inquirer*, OA 9450, Meese Files, Reagan Library.

6 26 October 1983, statement on the Grenada rescue operation, Box 1, Katherine Anderson Files, Reagan Library.

7 20 October 1983, NSC Records 1982–85, Box 42, Executive Secretariat Files, Reagan Library.

8 22 July 1983, letter from Higgins to Mr Ortega, Records, Series I, Box 17, AVH-271, Office of Correspondence Files, Reagan Library. See also 20 October 1983, Diplomatic cable from Secretary of State to US Embassy Bridgetown, NSC Records 1982–85, Box 42, Executive Secretariat Files, Reagan Library.

9 Undated, "On Current Strategy," Clark, NSC Agency Files Records, ID# 155209, Executive Secretariat Files, Reagan Library.

10 Formal letters signed by AVH (AVH 2700–AVH 2701), Record Series I, Box 17, Office of Correspondence Files, Reagan Library.

11 27 October 1983, NSC, National security affairs: Office of the Assistant to the President, October 1983, Executive Secretariat Files, Reagan Library.

12 2 November 1983, memo for the Senate, FG GOT-ORG, FG001, 111672–119734, Box 55, President of the US Files, Reagan Library.

13 Undated, NSC, Country File, Latin America, Falklands War, Grenada, Box 30, Executive Secretariat files, Reagan Library.

14 Weber, *Faking It*; Crandall, *Gunboat Diplomacy*.

15 Of course, not all the media or representatives in congress were opposed to the invasion, but for the stake of the argument I will concentrate on those that were.

16 28 August 1982, letter from Congressional Black Caucus to Reagan, Greece 280000–C0058 Grenada 201499, Box 80, CO Country Files, Reagan Library.

17 October 1983, major newspapers and Democrat comments, White House Office Records, OA 10495, 10446, 10047, Public Affairs Office Files, Reagan Library.

18 Ibid.

19 Ibid.

20 26 October 1983, "An Easy But Costly Invasion," *Hartford Courant*, OA 9450, Meese Files, Reagan Library; October 1983, major newspapers and Democrat

comments, White House Office Records, OA 10495, 10446, 10047, Public Affairs Office Files, Reagan Library.

21 27 October 1983, transcript of comments about the address by the President, OA 11244, Box 10, Barooday Files, Reagan Library.

22 Ibid.

23 Burrowes, *Revolution and Rescue Grenada*, 127.

24 February 1983, CHRON Files, ID# 123660, McFarlane Files, Reagan Library.

25 No. 38 to NSC, CIA and JCS, Box 90499-90500, Tillman Files, Reagan Library.

26 The variables used to produce these scores were stage, Reagan, secretaries of state, and advisors.

27 26 October 1983, White House talking points, Box 1, Anderson Files, Reagan Library.

28 Undated, background on legal-political rationale for US action, Records Office, Box 14, CFOA 1175, Counsel to the President, Reagan Library.

29 Ibid.

30 18 January 1983, Box 5, Sapia-Bosch Files, Reagan Library.

31 Whether it was widely shared is another question, for which the data does not cover.

32 13 February 1983, "The Strategic and Economic Importance of the Caribbean Sea Lanes," Office of Public Liaison, Series I, General Subject File OAS 12/51, Holladay Files, Reagan Library.

33 Crandall, *Gunboat Diplomacy*.

34 25 October 1983, statement by the president, NSC, Country File, Latin America: Falklands, Grenada, Box 30, Executive Secretariat Files, Reagan Library.

35 28 October 1983, Records Office, CFOA 1175, Box 14, Counsel to the President Files, Reagan Library.

36 21 October 1983, cable to Secretary of State, NSC Cable File Records 1982–85, Box 42, Executive Secretariat Files, Reagan Library.

37 20 October 1983, cable from US Embassy in Bridgetown to Secretary of State, NSC Cable File Records 1982–85, Box 42, Executive Secretariat Files, Reagan Library.

38 21 October 1983, contingency planning, National Security Decision Directives, ID #109126, Box 10, Executive Secretariat Files, Reagan Library.

39 21 October 1983, 3rd cable from US Embassy in Kingston to Secretary of State, NSC Cable File Records 1982–85, Box 42, Executive Secretariat Files, Reagan Library.

40 Grenada audio transcript, tape 2 side A TRT 10h41, Situation Room, White House Records, Reagan Library.

41 Adkin, *Urgent Fury*, 108.

42 20 October 1983, cable from US Embassy in Bridgetown to Secretary of State, NSC Cable File Records 1982–85, Box 42, Executive Secretariat Files, Reagan Library.

43 27 October 1983, address by the President to the nation, Records Office, CFOA 1175, 1176, Box 15, Counsel to the President Files, Reagan Library.

44 9 November 1983, memo for Senate committee, CO 058 Greece 280000–CO 058 Grenada 201499, Box 80, CO Country Files, Reagan Library.

45 12 December 1983, ibid.

46 Weber, *Faking It.*

47 Ducat, *The Wimp Factor.*

48 Weber, *Faking It*, 64.

49 27 October 1983, Records Office, CFOA 1175, 1176, Box 15, Counsel to the President, Reagan Library.

50 25 October 1983, statement by the President, National Security Affairs, Office of the Assistant to the President, Executive Secretariat Files, Reagan Library.

51 2 December 1983, Grenada students visit the White House, File, series XVIII, Grenada general correspondence, Box 96, Higgins Files, Reagan Library.

52 Weber, *Faking It*, 63.

53 3 January 1983, report of press view, memo to Craig Fuller from Dave Gergen, FG GOT-ORG FG001, ID #119528, Box 45, President of the US Files, Reagan Library. This memo compiles quotes from the press and, as such, quotes found in the document are not from the author but from columnists and congress.

54 26 October 1983, White House news summary, *New York Post*, OA 9450, Meese Files, Reagan Library.

55 3 January 1983, report of press view, memo from Gergen to Fuller, FG GOT–ORG FG001, ID #119528, Box 45, President of the US Files, Reagan Library.

56 3 January 1983, both quotations from a report of press view, memo from Gergen to Fuller, FG GOT–ORG FG001, ID #119528, Box 45, President of the US Files, Reagan Library.

57 Ibid.

58 27 October 1983, letter, Series VI: mail samples, Box 75, Higgins Files, Reagan Library.

59 28 October 1983, letter to the President from AM John Devis Lodge, CHRON File, RAC Box 6, McFarlane Files, Reagan Library; see also White House talking points, 26 October 1983, OA 18290, Linas Kojelis Files, Reagan Library.

60 28 October 1983, cable from US Embassy in Canberra to White House, National Security Affairs, Office of the Assistant to the President, Executive Secretariat Files, Reagan Library.

61 26 October 1983, White House news summary, comments in response to *New York Post*, OA 9450, Meese Files, Reagan Library.

62 Schweitzer, *Reagan's War*, 208.

63 26 October 1983, Series VI: mail samples, Box 75, Higgins Files, Reagan Library.

64 19 August 1983, memo to Clark from Charles Hill, Box 5, Sapia-Bosch Files, Reagan Library.

65 18 October 1983, memo to the President, NSC Agency File Records, RAC Box 6A, Executive Secretariat Files, Reagan Library.

66 18 November 1983, cable from Secretary of State to all US diplomatic posts (reminder of chronology of events), NSC Cables File Records 1982–85, Box 43, Executive Secretariat Files, Reagan Library.

67 25 October 1983, letter to congress by the president, Records Office, CFOA 1175, Box 14, Counsel to the President Files, Reagan Library.

68 27 October 1983, address by the president to the nation, Records Office, CFOA 1175- 1776, Box 15, Counsel to the President Files, Reagan Library.

69 24 October 1983, meeting with congressional leadership 8–9 pm, Country File, Latin America, Grenada, Box 3, Folder 1: 6/17/83–10/25/83, Executive Secretariat Files, Reagan Library.

70 25 October 1983, cabinet meeting agenda, Records Office, CFOA 1175, Box 14, Counsel to the President files, Reagan Library. Many other references can be found in the following files: National Security and Defence Directives, Executive Secretariat Files; Kojelis Files; National Security Council, Country File, Latin America, Grenada, Executive Secretariat Files.

71 24 February 1983, note to Dick Morris from Sapia-Bosch, Box 5, Sapia-Bosch Files, Reagan Library.

72 13 October 1983, National Security Affairs, Office of the Assistant to the President, Executive Secretariat Files, Reagan Library. See also Bureau of Intelligence and Research, current analysis, Series 12, August 1981, Subject File, Series I, RAC Box 8, Fontaine Files, on how Bishop was dealing with political opponents and the fate he reserved for them.

73 3 June 1983, cable from US Embassy in Bridgetown to Secretary of State, Box 5, Sapia-Bosch Files, Reagan Library.

74 25 October 1983, cable from Secretary of State to US Embassy Ottawa, NSC Cable File Records 1982–85, Box 42, Executive Secretariat Files, Reagan Library.

75 25 October 1983, cable from Secretary of State to US Embassy Caracas, NSC Cable File Records 1982–85, Box 42, Executive Secretariat files, Reagan Library.

76 Adkin, *Urgent Fury*.

77 28 October 1983, White House to Vice-President George H. Bush, NSC Country File, Latin America, Falklands War, Grenada, Box 30, Executive Secretariat Files, Reagan Library.

78 28 October 1983, polls and opinions, OA11244, Box 10, Barooday Files, Reagan Library.

79 Grenada general correspondence, Series VI, Box 75, Higgins Files, Reagan Library.

80 1 June 1983, Country file, Latin America, Grenada, Box 3, Folder 1 6/17/83–10/25/83, Executive Secretariat Files, Reagan Library.

81 Undated, talking points, NSC Country File: Latin America, Falklands War, Grenada, Box 30, Executive Secretariat Files, Reagan Library.

82 20 October 1983, NSC Country File: Latin America, Falklands War, Grenada, Box 30, Executive Secretariat Files, Reagan Library.

83 28 October 1983, basic facts, Grenada, Records Office, CFOA 1175, Box 14, Counsel to the President, Reagan Library. See also talking points for Reagan with special commissary McNeil, 24 October 1983, NSC Country File, Latin America, Falklands War, Grenada, Box 30, Executive Secretariat Files, Reagan Library.

Chapter Nine

1 Jervis, *Perception and Misperception*; Paul et al., *Status in World Politics*; McCourt, *Britain and World Power*.

2 Ducat, *The Wimp Factor*; Weber, *Faking It*.

3 Dyson, *The Blair Identity*.

4 Lakoff, *The Political Mind: Why You Can't Understand 21st-Century*; *Moral Politics*; *The Political Mind: A Cognitive Scientists' Guide*.

5 Horowitz et al., *Why Leaders Fight*; Lakoff, *The Political Mind: Why You Can't Understand 21st-Century*; *Moral Politics*; *The Political Mind: A Cognitive Scientists' Guide*; Dyson, *The Blair Identity*; Yahri-Milo, *Who Fights for Reputation*.

6 Volgy et al., *Major Powers and the Quest for Status*; Paul et al., *Status in World Politics*.

7 Suzuki, "Seeking 'Legitimate' Great Power Status"; Lebow, *Why Nations Fight*; Wolf, "Respect and Disrespect in International Politics."

8 Greve and Levy, "Power Transition, Status Dissatisfaction and War."

9 This study is based on preliminary results. The bigger project that aims at comparing the Russian invasion of Ukraine to the Chinese policy in the South China Sea and how these two decisions challenge the current international order and attempt to make a new world emerge, will be presented at two international conferences in 2023: the 27th Annual World Convention of the Association for the Study of Nationalities, Columbia University; and the conference "Rethinking the Past and Present of Liberal Internationalism" organized by the London School of Economics and the University City of London.

10 *Chargé d'affaires* Sun Xianggang's press conference, Chinese embassy in the Philippines, 1 April 2014. See also: Foreign ministry spokesperson Hua Chunying's regular press conference, 8 September 2016: "China will never accept any third-party settlement adopted without China's consent or any solution shoved under its nose."

11 Ambassador to Belgium Qu Xing's press conference, 15 April 2015: "The difference between the two is that the former (China) wins people's support by virtue, while the latter (the US) forces people to submit."

12 Foreign ministry spokesperson Hua Chunying's remarks on the conclusion of the hearing on issues relating to jurisdiction and admissibility by South China Sea arbitral tribunal established at the request of the Philippines, 14 July 2015. See also Foreign Ministry spokesperson Geng Shuang's regular press conference, 17 February 2017: "China is not known for bullying the weak. China has only been commended for speaking out for developing countries in defiance of the bullying by strong powers."

13 A representative example can be found in Foreign Ministry spokesperson Hua Chunying's regular press conference, 29 April 2016: "The US took a preemptive move and introduced the freedom of navigation operations in 1979 before the signing UN Convention on the Laws of the Sea (UNCLOS), with the purpose of ensuring the maximum freedom and maneuverability of US military forces in all the oceans. It lays bare the US attempt to dominate the order for the seas and oceans, its hegemonic logic of complying with international law only when the law serves its interests and US enjoying privileges."

14 A representative example of such discursive constructions can be found in the 26 April 2016 press conference aimed at the US military: "stop playing up the issue of the South China Sea, stop hyping up tensions and stop showing off military might in the South China Sea."

15 Foreign ministry spokesperson Hong Lei's regular press conference, 20 March 2016.

16 Foreign ministry spokesperson Hua Chunying's remarks on the Philippines' playing up and airing of a documentary on the South China Sea issue, 29 June 2015.

17 Lowenheim and Heimann, "Revenge in International Politics."

18 Foreign Ministry Spokesperson Lu Kang's Remarks on the US Statement about Issues Relating to the Arbitration Unilaterally Initiated by the Philippines, 24 July 2015: "the US side just acts like an 'arbitrator outside the tribunal,' designating the direction for the arbitral tribunal … This is inconsistent with the position the US side claims to uphold on issues concerning the South China Sea disputes."

19 See, for example: Foreign ministry spokesperson Hua Chunying's regular press conference, 5 January 2016.

20 See, for example: Foreign ministry spokesperson Hong Lei's regular press conference, 20 January 2016.

21 Salter, *Carl Schmitt*; Boyle, "The Coming Illiberal Order"; Specter, "Grossraum and Geopolitics."

22 It is important to keep this in mind: in official Russian discourses, Ukraine is never described as the main villain. If Ukraine is a villain, it is because of the action of

others. As such, discourses always associate Ukraine to the actions of NATO or the US, France, and Germany. When the blame is put on Ukraine, Russia never refers to the country but to its leaders. This strategy has a double discursive objective: first, it helps build the narrative that Ukrainians are victims, too, and should support Russia; second, it helps develop a nurturant parent tone. The same strategy is used by China towards the Philippines in regard to the South China Sea.

23 Vladmir Putin, address by the president of the Russian Federation, 24 February 2022. See also Sergei Lavrov interview, TASS, 27 December 2022.

24 Vladmir Putin, address by the president of the Russian Federation, 24 February 2022.

25 Dmitri Medvedev, 19 January 2017. The feeling not to be considered properly by a Western country is a leitmotiv in Russian discourses since the end of the Cold War. This narrative takes its source from the Russian veto of Kosovo in 1998 being ignored by the US, and the false promises that the US would not expand NATO after the Cold War. An example can be found in another Medvedev discourse: "there came a time when I thought we were friends … and that we understood each other well. There is nothing left of this trust now." Interview for Handeslblatt, 11 February 2016.

26 Sergei Lavrov interview for TASS, 27 December 2020: "who [Ukrainian government] decided to strip the Russian language of its special status and talked about annihilating Russia." See also Vladimir's Putin discourse on the annexation of the 4 regions of Ukraine, 20 September 2022.

27 See for example: Briefing by foreign ministry spokesperson Maria Zakharova, 22 April 2021.

28 For a representative example see Sergei Lavrov's news conference on European security issues, Moscow, 1 December 2022.

29 Briefing by Foreign ministry spokesperson Maria Zahkarova, 1 October 2020.

30 Dmitri Medvedev interview with CNBC, 15 October 2014: "It is sad, it's like some kind of mental aberration … President Obama said in an address at the UN that the threats and challenges facing humanity are, in this particular order: The Ebola virus, the Russian federation."

31 Sergei Lavrov interview with Trend newspaper, 21 August 2020: "we hope that a sober analysis of the multipolar world will naturally prompt the EU to rethink its obviously outdated approaches." See also Medvedev's criticisms of amateurism dating from 2014: "serious politicians should choose their words carefully"; Interview, 15 October 2014.

32 Vladimir Putin, discourse on the annexation of the 4 regions of Ukraine, 20 September 2022: "They imagined that they could once more put the world at their feet. However, this prospect was far from exciting to all but the hardcore masochists and practitioners of other non-traditional forms of international relations."

33 Vladimir Putin, interview with the chiefs of federal governments, 16 March 2022.

34 Briefing by Foreign ministry spokesperson Maria Zahkarova, 22 April 2021.

35 Foreign ministry spokesperson Maria Zahkarova interview with director general of Rossiya Segodnya international information agency Dmitry Kisolov: "He [Volodymyr Zelensky] is ready to pay any price, such as pandering to neo-Nazis and ultra-radicals."

36 For a representative example see Dmitri Medvedev, 15 October 2014: "Despite all the current difficulties (political situation and the fact Ukraine cannot afford to pay), Russia undertook certain obligations under this contract and will fully meet them."

37 Vladimir Putin answered questions on the article "On the historical unity of Russians and Ukrainians," 13 July 2021: "We are not dragging anyone there. I repeat, we are looking for compromise solution that create a situation that our partners feel satisfied with."

38 See, for example, Sergei Lavrov discourse, 1 December 2022: "we are using precision weapons to knock out of service the energy facility that are important for the operation of the Ukrainian armed forces." See also Vladimir Putin discourse on 16 March 2022: "soldiers and officers … do all what is in their power to avoid civilian casualties in Ukrainian cities."

39 Sergei Lavrov discourse on 31 December 2021.

40 Vladimir Putin, address by the president of the Russian federation, 24 February 2022: "There is no alternative to the Minsk package of measures."

41 To strengthen the comparison, Russian officials qualify the economic sanctions taken by Occidental countries as a "blitzkrieg." See Sergei Lavrov interview for the movie "Nazism under investigation," 26 November 2022.

42 Vladimir Putin, address by the president of the Russian Federation, 24 February 2022.

43 Vladimir Putin, meeting with permanent members of the United Nations Security Council, 10 October 2022.

BIBLIOGRAPHY

Abdelal, R., Y. Herrera, A.I. Johnston, and R. McDarmott, eds. *Measuring Identity: A Guide for Social Scientists*. Cambridge: Cambridge University Press, 2009.

Adkin, M. *Urgent Fury the Battle for Grenada: The Truth behind the Last US Military Operation since Vietnam*. London: Leo Cooper, 1989.

Ahmed, S. *The Cultural Politics of Emotions*. Edinburgh: Edinburgh University Press, 2014.

Aklaev, A.R. *Etnopoliticheskaia Konflitktologiia: Analiz i Menedzhment: Uchebnoe Pocobie [Ethnopolitical Conflict Studies: Analysis and Management: Textbook]*. Moscow: Delo, 2005.

Albaret, M., and D. Placidi-Frot. "Les petits Etats au Conseil de sécurité : des strapontins à l'avant-scène." *Critique Internationale* 71, no. 2 (2016): 19–38.

Allès, D. "Ethnic and Religious Violence." In *Routledge Handbook of Security Studies*, edited by M.D. Cavelty and T. Balzacq, 193–203. London: Routledge, 2017.

Allison, G. *Destined for War: Can America and China Escape Thucydides' Trap?* Boston: Houghton Mifflin Harcourt, 2017.

Anno, S.E., and W.E. Einspach. "Command and Control and Communications Lessons Learned: Iranian Rescue, Falklands Conflict, Grenada Invasion, Libya Raid." Alabama: Air War College Research Report, 1988.

Applegate, M.L. *The Egyptian Book of Life: Symbolism of Ancient Egypt Temple and Tomb Art*. Deerfield Beach: Health Communications Inc., 2001.

Aragones, E., I. Gilboa, A. Postlewaite, and D. Schmeidler. "Rhetoric and Analogies." *Research in Economica* 68 (2013): 1–10.

Aron, R. *Paix et guerre entre les nations*. Paris : Calmann-Lévy, 1962 (2004).

Avey, P. *Tempting Fate: Why Nonnuclear States Confront Nuclear Opponents*. Ithaca: Cornell University Press, 2019.

Badie, B. "French Power-Seeking and Overachievement." In *Major Powers and the Quest for Status in International Politics: Global and Regional Perspectives*, edited by T.J. Volgy, R. Corbetta, K.A. Grant, and R.G. Baird, 97–144. New York: Palgrave Macmillan, 2011.

– *L'impuissance de la puissance*. Paris : CNRS Editions, 2013.

Baillat, A., F. Emprin, and F. Ramel. "Des mots et des discours, du quantitatif au qualitatif." In *Méthodes de recherche en relations internationales*, edited by G. Devin, 227–46. Paris : SciencesPo les Presses, 2016.

Baranets, E. *Comment Perdre une Guerre: Une Théorie du Contournement Démocratique*. Paris : CNRS Editions, 2017.

Barkin, S. *Realist Constructivism: Rethinking International Relations Theory*. Cambridge: Cambridge University Press, 2010.

Barnett, M., and R. Duvall. "Power in International Politics." *International Organization* 59, no. 1 (2005): 39–75.

Barnhart, J. "Prestige, Humiliation and International Politics." PhD dissertation, University of California, Los Angeles, 2013.

Baron, J. *Great Power Peace and American Primacy: The Origins and Future of a New International Order*. London: Palgrave Macmillan, 2013.

Bell, D. *Men on Horseback: The Power of Charisma in the Age of Revolution*. New York: Farrar, Strauss and Giroux, 2020.

Bellot, A. "War Speeches and Nationalist Discourse in the Malvinas/Falklands War." In *At a Time of Crisis: English and American Studies in Spain*, edited by M. Aligne, M. Meyer, E. Pladevall, and S. Tubau. Barcelona: AEDEAN, 2012.

Bergmane, U. "La consultation des archives des Affaires étrangères. Le cas français et américain." In *Méthodes de recherche en relations internationales*, edited by G. Devin, 59–70. Paris : SciencesPo les Presses, 2016.

Blechman, B., and R. Weinland. "Why Coaling Stations Are Necessary in the Nuclear Age." *International Security* 2, no. 1 (1977): 88–99.

Bleiker, R., and A. Kay. "Representing HIV/AIDS in Africa." *International Studies Quarterly* 51, no. 1 (2007): 139–63.

Bluth, C. "The British Resort to Force in the Falklands/Malvinas Conflict 1982: International Law and Just War Theory." *Journal of Peace Research* 24, no. 1 (1987): 5–20.

Bobić, M. *Why Minor Powers Risk Wars with Major Powers: A Comparative Study of the Post-Cold War Era*. Bristol: Bristol University Press, 2019.

Bodinier, G., ed. *Le retour de la France en Indochine 1945–1946, textes et documents*. Vincennes : Service historique de l'Armée de Terre, 1987.

– ed. *Indochine 1947: Règlement politique ou solution militaire, textes et documents*. Vincennes : Service historique de l'Armée de Terre, 1989.

Boin, A., P. Hart, E. Stern, and B. Sundelius. *The Politics of Crisis Management: Public Leadership under Pressure*. Cambridge: Cambridge University Press, 2013.

Bonhême, M-A., and A. Forgeau. *Pharaon: les secrets du pouvoir*. Paris : Armand Colin, 1988.

Boulding, K. "National Images and International Systems." *Journal of Conflict Resolution* 3, no. 2 (1971): 120–31.

Bowman, J. *Honor: A History*. New York: Encounter Books, 2006.

Boyle, J. "The Coming Illiberal Order." *Survival* 58, no. 2 (2016): 35–66.

Bramley, V. *Excursion to Hell: Mount Longdon*. London: Pan Books, 1992.

Breuning, M. *Foreign Policy Analysis: A Comparative Introduction*. New York: Palgrave Macmillan, 2007.

Brice, B. "Un président en colère. Eléments de méthode pour une analyse émotionnelle à partir d'un texte de mémoires." Workshop: Vers une synergie des méthodes des relations internationales et de la sociologie des mobilisations. Le cas des émotions dans les mobilisations violentes. Paris : Congrès Association Française de Science Politique, 2013.

Brown, C. "Do Great Powers Have Great Responsibilties? Great Powers and Moral Agency." *Global Society* 18, no. 1 (2004): 5–19.

Buckley, S. *The Prime Minister and Cabinet*. Edinburgh: Edinburgh University Press, 2006.

Bull, H. *The Anarchical Society: A Study of Order in World Politics*. Basingstoke: Macmillan, 1977.

Burbank, J., and F. Cooper. *Empires: De la Chine ancienne à nos jours*. Paris : Editions Payot & Rivages, 2011.

Burrowes, R. *Revolution and Rescue Grenada: An Account of the U.S.-Caribbean Invasion*. New York: Greenwood Press, 1988.

Callahan, W.A. "National Insecurities: Humiliations, Salvation, and Chinese Nationalism." *Alternatives* 29, no. 2 (2004): 199–218.

Calvert, P. "Sovereignty and the Falklands Crisis." *International Affairs* 59, no. 3 (1983): 405–13.

Campbell, J. *Margaret Thatcher. Volume One: The Grocer's Daughter*. London: Jonathan Cape, 2001.

Carrère d'Encausse, H. *Six années qui ont changé le monde 1985–1991: La chute de L'Empire soviétique*. Paris : Fayard, 2015.

Cartledge, P. *Sparta and Lakonia: A Regional History 1300 to 362 BC*. London: Routledge, 2001 (1979).

Chassaigne, P. *La Grande-Bretagne et le monde de 1815 à nos jours*. Paris: Armand Colin, 2009.

Chiozza, G., and A. Choi. "Guess Who Did What: Political Leaders and the Management of Territorial Disputes, 1950–1990." *Journal of Conflict Resolution* 47, no. 3 (2003): 251–78.

Chiozza, G., and H.E. Goemens. *Leaders and International Conflict*. Cambridge: Cambridge University Press, 2011.

Clarke, J. *Operation Urgent Fury: The Invasion of Grenada, October 1983*. Washington, DC: US Government Printing Office, 2018.

Clément, M. "Al-Muhajiraun in the United Kingdom: The Role of International Non Recognition in Heightened Radicalization Dynamics." *Global Discourse: An Interdisciplinary Journal of Current Affairs* 4, no. 4 (2014): 428–43.

Clément, M., T. Lindemann, and E. Sangar. "The 'Hero-Protector Narrative': Manufacturing Emotional Consent for the Use of Force." *Political Psychology* 38, no. 6 (2017): 991–1008.

Cole, R.H. *Operation Urgent Fury Grenada*. Washington, DC: Joint History Office, Office of the Chairman of the Joint Chiefs of Staff, 1997.

Congyan, C. "New Great Power and International Law in the 21st Century." *European Journal of International Law* 24, no. 3 (2013): 755–95.

Connell, R. *Masculinities*. Berkeley: University of California Press, 2005.

Corbin, A., J-J. Courtine, and G. Vigarello, eds. *A History of Virility*. New York: Columbia University Press, 2017.

Cornell-Smith, G. "The Grenada Invasion in Historic Perspective: From Monroe to Reagan." *Third World Quarterly* 6, no. 2 (1984): 432–45.

Cosme, P. *Auguste*. Paris : Perrin, 2005.

Courtois, S., ed. *Ernst Nolte: Fascisme et Totalitarisme*. Paris : Editions Robert Laffont, 2008.

Crandall, R. *Gunboat Diplomacy: U.S. Intervention in the Dominican Republic, Grenada and Panama*. New York: Rowman and Littlefield, 2006.

Crawford, N. "The Passion of World Politics." *International Society* 24, no. 4 (2000): 120–2.

Dalloz, J. *La guerre d'Indochine 1945–1954*. Paris : Editions du Seuil, 1987.

Darwin, J. *The Empire Project: The Rise and Fall of the British World-System, 1830–1970*. Cambridge: Cambridge University Press, 2009.

Davis, J.W. *Threats and Promises: The Pursuit of International Influence*. Baltimore: Johns Hopkins University Press, 2000.

Deng, Y. *China's Struggle for Status: The Realignment of International Relations*. Cambridge: Cambridge University Press, 2008.

Deriglazova, L. *Great Powers, Small Wars: Asymmetric Conflict Since 1945*. Baltimore: Johns Hopkins University Press, 2014.

de Larrinaga, M., and M.G. Doucet. "Governmentality, Soverign Power and Intervention. Security Council Resolution and the Invasion of Iraq." In *Security and Global Governmentality: Globalization, Governance and the State*, edited by M. de Larrinaga and M.G. Doucet, 96–110. New York: Routledge, 2010.

de Waal-Andrews, W., A.P. Gregg, and J. Lammers. "When Status Is Grabbed and When Status Is Granted: Getting Ahead in Dominance and Prestige Hierarchies." *British Journal of Social Psychology* 54 (2015): 445–64.

de Wailly, H. *1945 L'Empire Rompu: Syrie, Algérie, Indochine*. Paris : Perrin, 2012.

Donaghy, A. *The British Government and the Falklands Islands 1974–1979*. New York: Palgrave Macmillan, 2014.

Dodds, K. *Pink Ice: Britain and the South Atlantic Empire*. London: LB. Tauris, 2002.

Doran, C.F., K.Q. Hill, and K. Mladenka. "Threat, Status Desequilibrium, and National Power." *British Journal of International Studies* 5, no. 1 (1979): 37–58.

Ducat, J. *The Wimp Factor: Gender Gaps, Holy Wars, and the Politics of Anxious Masculinity*. Boston: Beacon Press, 2004.

Dufour, J-L. *Un siècle de crises internationales: De Pékin (1900) au Caucase (2008)*. Bruxelles : André Versailles Editeur, 2009.

Duque, M. "Recognizing International Status: A Network Analysis of Diplomatic Representation." Paper presented at ISA Atlanta 2016, 2015.

Dyson, S.B. *The Blair Identity: Leadership and Foreign Policy*. Manchester: Manchester University Press, 2009.

Elgy, E. *Histoire de la IVe République: De 1945 à 1957 La modernisation de la France, l'Europe en marche, la guerre d'Indochine*. Paris: Robert Laffont, 2018 (1965).

Ellerby, C. "The Role of the Falkland Lobby, 1968–1992." In *International Perspectives on the Falklands Conflict*, edited by A. Danchev, 85–108. London: Palgrave Macmillan, 1992.

Evans, G., and M. Sahnoun. *The Responsibility to Protect: Research, Bibliography, Background*. Ottawa: International Development Research Centre, 2001.

Falkland Islands Government. *Results of the Referendum on the Political Status of the Falkland Islands*. [Press release], 12 March 2013. Available at http://www.falklands.gov.fk/results-of-the-referendum-on-the-political-status-of-the-falkland-islands/.

Fairclough, N. *Critical Discourse Analysis: The Critical Study of Language*. Harlow: Pearson Education Limited, 2010.

Ferrari, P., and J.M. Verret. *Une guerre sans fin, Indochine 1945–1954*. Paris : Lauvauzelle, 1984.

Fierke, K.M. "Agents of Death. The Structural Logic of Suicide Terrorism and Martyrdom." *International Theory* 1, no. 1 (2009): 155–84.

– *Diplomatic Interventions: Conflict and Change in a Globalizing World*. London: Palgrave Macmillan, 2005.

– *Political Self-Sacrifice: Agency, Body and Emotion in International Relations*. Cambridge: Cambridge University Press, 2013.

Fikenscher, S-V., L. Jaschob, and R. Wolf. "Seeking Status Recognition through Military Symbols: German and Indian Armament Policies between Strategic Rationalizations and Prestige Motives." In *Recognition in International Relations: Rethinking a Political Concept in a Global Context*, edited by C. Daase, C. Fehl, A. Geis, and G. Kolliarakis, 86–103. London: Palgrave Macmillan, 2015.

Finnemore, M. *The Purpose of Intervention: Changing Beliefs about the Use of Force*. London: Cornell University Press, 2003.

Fischerkeller, M. "David versus Goliath: Cultural Judgements in Asymmetric Wars." *Security Studies* 7, no. 4 (1998): 1–43.

Forsberg, T., R. Heller, and R. Wolf. "Russia and the Quest for Status." *Communist and Post-Communist Studies* 47 (2014): 261–8.

Foxall, A. "Photographing Vladimir Putin: Masculinity, Nationalism and Visuality in Russian Political Culture." *Geopolitics* 18 (2013): 132–56.

Franchini, P. *Les Guerres d'Indochine: De la conquête française à 1949*. Paris : Tallandier, 2011.

Franck, T.M. "Dulce et Decorum Est: The Strategic Role of Legal Principles in the Falklands War." *American Journal of International Relations* 77, no. 11 (1983): 109–24.

Fravel, M.T. "The Limits of Diversion: Rethinking Internal and External Conflict." *Security Studies* 19, no. 2 (2010): 307–41.

Freedman, L. *Britain and the Falklands War (Making Contemporary Britain)*. Oxford: Wiley Blackwell, 1988.

– "The War of the Falkland Islands." *Foreign Affairs* 61, no. 1 (1982): 196–210.

Freedman, L., and V. Gamba. *Signals of War: The Falklands Conflict of 1982*. Princeton: Princeton University Press, 1991.

Frémeaux, J. *De quoi fut fait l'empire? Les guerres coloniales au XIX siècle*. Paris : CNRS Editions, 2010.

Friedberg, A. *The Weary Titan: Britain and the Experience of Relative Decline 1895–1905*. Princeton: Princeton University Press, 1988.

Galtung, J. "A Structural Theory of Imperialism." *Journal of Peace Research* 8, no. 2 (1971): 81–117.

Garnett, M. "Banality and Politics: Margaret Thatcher and the Biographers." *Political Studies Review* 5 (2007): 172–82.

Geiss, I. *Geschichte im Überblick: Daten, Fakten und Zusammenhänger der Weltgeschichte*. Köln: Anaconda Verlag GmbH, 2008.

Gibran, D.K. *The Falklands War: Britain Versus the Past in the South Atlantic*. Jefferson: McFarland & Company, 1998.

Gilmore, W. *The Grenada Intervention: Analysis and Documentation*. New York: Facts on File Inc., 1984.

Gilpin, R. *War and Change in World Politics*. Cambridge: Cambridge University Press, 1981.

Go, J. "American Decline and Performative War or How to Do Things with Force." In *Coping with Geopolitical Decline: The United States in European Perspective*, edited by F. Mérand, 163–93. Montreal: McGill-Queen's University Press, 2020.

Gochman, C.S., and Z. Maoz. "Militarized Interstate Disputes, 1816–1976: Procedures, Patterns, and Insights." *Journal of Conflict Resolution* 28, no. 4 (1984): 585–616.

Goldstein, J.S. *War and Gender*. Cambridge: Cambridge University Press, 2001.

Goscilo, H., ed. *Putin as Celebrity and Cultural Icon*. New York: Routledge, 2013.

Grandpierron, M. "The British Use of Domestic Political Situations during the Falklands War." *Round Table* 111, no. 1 (2022): 56–68.

– "Preserving 'Great Power Status': The Complex Case of the British Intervention in the Falklands (1982)." *Croatian International Relations Review* 23, no. 79 (2017): 127–56.

Granvaud, R. *Areva en Afrique: Une face cachée du nucléaire français.* Marseille : Agone, 2012.

Gregory, T. "Photographing War: Don McCullin, Vietnam and the Politics of Emotion." In *Politics and War*, edited by L. Ahäll and T. Gregory, 182–94. London: Routledge, 2015.

Greve, A.Q., and J.S. Levy. "Power Transition, Status Dissatisfaction, and War: The Sino Japanese War of 1894–1895." *Security Studies* 1 (2018): 148–78.

Grimal, H. *La décolonisation de 1919 à nos jours.* Bruxelles : Editions Complexe, 1985.

Gros, Y. *Histoire de la guerre d'Indochine.* Paris : Plon, 1979.

Gueniffey, P., and T. Lentz, eds. *La fin des empires.* Paris : Perrin, 2016.

Gustafson, L.S. *The Sovereignty Dispute over the Falkland (Malvinas) Islands.* Oxford: Oxford University Press, 1988.

Haass, R. *Intervention: The Use of American Military Force in the Post-Cold War World.* Washington, DC: The Brookings Institution, 1999.

Hall, T.H. *Emotional Diplomacy: Official Emotion on the International Stage.* Ithaca: Cornell University Press, 2015.

Halperin, S., and O. Heath. *Political Research: Methods and Practical Skills.* 2nd ed. Oxford: Oxford University Press, 2017.

Hamilton-Hart, N. *Hard Interests, Soft Illusions: Southeast Asia and American Power.* Ithaca: Cornell University Press, 2012.

Hanink, J. *How to Think about War: An Ancient Guide to Foreign Policy.* Princeton: Princeton University Press, 2019.

Harkavy, R.E. *Great Power Competition for Overseas Bases: The Geopolitics of Access Diplomacy.* Oxford: Pergamon Press, 1982.

Hastings, M., and S. Jenkins. *The Battle for the Falklands.* London: Pan Macmillan, 2010.

Hayton, B. *The South China Sea: The Struggle for Power in Asia.* New Haven: Yale University Press, 2014.

Heller, R. "Defending Social Status: Why Russia's Ukraine Policy Is More than Regional Leadership." *Rising Power Quarterly* 3, no. 1 (2018): 137–59.

– "Russia's Power Policies towards Ukraine: Social Status Concerns and the Role of Emotions." In *Russia in the Changing International System*, edited by D.E. Larlar and E. Ersen, 169–86. London: Palgrave Macmillan, 2019.

– "Russia's Quest for Respect in the International Conflict Management in Kosovo." *Communist and Post-Communist Studies* 47, nos 3–4 (2014): 333–43.

Henchan, M.T., and J. Vasquez. "The Changing Probability of Interstate War, 1816–1992." In *The Waning of Major War: Theories and Debates*, edited by R. Väyrynen, 280–99. London: Routledge, 2006.

Hermann, M.G. "Assessing Foreign Policy Role Orientation of Sub-Saharan African Leaders." In *Role Theory and Foreign Policy Analysis*, edited by S.G. Walker, 161–98. Durham: Duke University Press, 1987.

– "Assessing Leadership Styles: Trait Analysis." In *The Psychological Assessment of Political Leaders*, edited by J.M. Post, 178–214. Ann Arbor: University of Michigan Press, 2003.

– "How Decision Units Shape Foreign Policy: A Theoretical Framework." *International Studies Review* 3, no. 2 (2001): 47–81.

Hill, F., and C. Gaddy. *Mr Putin: Operative in the Kremlin*. Washington, DC: Brookings Institution, 2012.

Hobbes, T. *The Leviathan*. Cambridge: Cambridge University Press, 1996 (1651).

Hoffman, F.L., and O.M. Hoffman. *Sovereignty in Dispute: The Falklands/Malvinas, 1493–1982*. Boulder, CO: Westview Press, 1984.

Holleindre, J-V. "A Certain Idea of Grandeur: French Military Interventionism and Postcolonial Responsibility." In *The Ethics of War and Peace Revisited*, edited by D.R. Brunstetter and J-V. Holeindre, 139–58. Washington, DC: Georgetown University Press, 2018.

Holsti, K.J. *The State, War, and the State of War*. Cambridge: Cambridge University Press, 1996.

Horowitz, M.C., A.C. Stam, and C.M. Ellis. *Why Leaders Fight*. Cambridge: Cambridge University Press, 2015.

Huberman, B.A., C.H. Loch, and A. Onculer. "Status as a Valued Resource." *Social Psychology Quarterly* 67, no. 1 (2004): 103–14.

Hudson, V.M. "Birth Order of World Leaders: An Explanatory Analysis of Effects on Personality and Behavior." *Political Psychology* 11, no. 3 (1990): 583–601.

Ikenberry, G.J. *Power, Order and Change in World Politics*. Cambridge: Cambridge University Press, 2014.

Jackson, R. *The Global Covenant: Human Conduct in a World of States*. Oxford: Oxford University Press, 2002.

Janis, I.L. *Groupthink: Psychological Studies of Policy Decisions and Fiascoes*. Boston: Houghton Mifflin, 1982.

Jervis, R. "Do Leaders Matter and How Would We Know?" *Security Studies* 22 (2013): 153–79.

– *How Statesmen Think: The Psychology of International Politics*. Princeton: Princeton University Press, 2017.

– *The Logic of Images in International Relations*. New York: Columbia University Press, 1989.

– "Perceiving and Coping with Threat." In *Psychology and Deterrence*, edited by R. Jervis, R.N. Lebow, and J.G. Stein, 13–33. Baltimore: Johns Hopkins University Press, 1985.

– *Perception and Misperception in International Politics*. Princeton: Princeton University Press, 1976.

Joyner, C.C. "Reflections on the Lawfulness of Invasion." *American Journal of International Law* 78, no. 1 (1984): 131–44.

Kantorowicz, E. *The King's Two Bodies*. Princeton: Princeton University Press, 1957.

Kagan, D. *A Twilight Struggle: American Power in Nicaragua, 1977–1990*. New York: Free Press, 1996.

Keegan, J. *A History of Warfare*. New York: Knopf, 1993.

Kertzer, J. *Resolve in International Politics*. Princeton: Princeton University Press, 2016.

Kinney, D. *National Interest/National Honor: The Diplomacy of the Falklands Crisis*. Westport: Praeger Publishers, 1990.

Kinne, B. "Dependent Diplomacy: Signaling, Strategy, and Prestige in the Diplomatic Network." *International Studies Quarterly* 58 (2014): 247–59.

Khong, Y.F. *Analogies at War, Korea, Munich, Dien Bien Phu, and the Vietnam Decisions of 1965*. Princeton: Princeton University Press, 1992.

Klotz, A., and D. Prakash, eds. *Qualitative Methods in International Relations: A Pluralist Guide*. New York: Palgrave Macmillan, 2008.

Lai, B., and D. Reiter. "Rally 'Round the Union Jack? Public Opinion and the Use of Force in the United Kingdom, 1948–2001." *International Studies Quarterly* 49, no. 2 (2005): 253–72.

Lakoff, G. *Moral Politics: How Liberals and Conservatives Think*. Chicago: University of Chicago Press, 2002.

– *The Political Mind: A Cognitive Scientist's Guide to Your Brain and Its Politics*. London: Penguin Books, 2009.

– *The Political Mind: Why You Can't Understand 21st-Century American Politics with an 18th-Century Brain*. London: Viking, 1996.

Landry, R. "L'analyse de contenu." In *Recherche sociale: De la problématique à la collecte des données,* 2nd ed., edited by B. Gauthier, 337–59. Saint-Foy : Presses de l'université du Québec, 1993.

Larson, D.W., and A. Schevchenko. "Russia Says No: Power, Status, and Emotions in Foreign Policy." *Communist and Post-Communist Studies* 47, nos 3–4 (2014): 269–79.

Laver, M., K. Benoit, and J. Garry. "Extracting Policy Positions from Political Texts Using Words as Data." *American Political Review* 97, no. 2 (2003): 311–31.

Le Courriard, D. "Les socialistes et les débuts de la guerre d'Indochine (1946–1947)." *Revue d'Histoire Moderne et Contemporaine* 31, no. 2 (1984): 334–53.

Lebow, R.N. "Miscalculation in the South-Atlantic: The Origins of the Falklands War." *Journals of Strategic Studies* 6, no. 1 (1983): 5–35.
– *Why Nations Fight*. Cambridge: Cambridge University Press, 2010.
Levy, J.S. "The Diversionary Theory of War: A Critique." In *Handbook of War Studies*, edited by M.I. Midlarsky, 259–88. Ann Arbor: University of Michigan Press, 1989.
– "Prospect Theory, Rational Choice, and International Relations." *International Studies Quarterly* 41, no. 1 (1997): 87–112.
– *War in the Modern Great Power System 1495–1975*. Lexington: University Press of Kentucky, 1983.
Levy, J.S., and L.I. Vakili. "Diversionary Action by Authoritarian Regimes: Argentina in the Falklands/Malvinas Case." In *The Internationalization of Communal Strife*, edited by M.I. Midlarsky, 118–48. New York: Routledge, 1992.
Lewis, P., G. Williams, and P. Clegg, eds. *Grenada: Revolution and Invasion*. Jamaica: The University of the West Indies Press, 2015.
Lindemann, T. *Causes of War: The Struggle for Recognition*. Colchester, UK: ECPR Press, 2010.
Lowenheim, O., and G. Heimann. "Revenge in International Politics." *Security Studies*, 17, no. 4 (2017): 685–724.
Lukes, S. *Power: A Radical View*. London: Macmillan Press, 1974.
MacDonald, P.K., and J.M. Parent. *Twilight of the Titans: Great Power Decline and Retrenchment*. Ithaca: Cornell University Press, 2018.
Mack, A. "Why Big Nations Lose Small Wars: The Politics of Asymmetric Conflict." *World Politics* 27, no. 2 (1975): 175–200.
Markey, D. "The Prestige Motive in International Relations." PhD dissertation, Princeton University, 2000.
Marsh, S. "Anglo-American Relations 1950–1951: Three Strikes for British Prestige." *Diplomacy and Statecraft* 23 (2012): 304–30.
Marshall, T. *Class, Citizenship and Social Development: Essays*. Chicago: University of Chicago Press, 1977.
Marx, R. *Histoire de l'Angleterre*. Paris : Fayard, 1993.
Matard-Bonucci, M-A., and P. Milza. *L'homme nouveau dans l'Europe fasciste (1922–1945): Entre dictature et totalitarisme.* Paris : Fayard, 2004.
McCourt, D.M. *Britain and World Power Since 1945: Constructing a Nation's Role in International Politics.* Ann Arbor: University of Michigan Press, 2014.
– "Role-Playing and Identity Affirmation in International Politics: Britain's Reinvasion of the Falklands, 1982." *Review of International Studies* 37, no. 4 (2011): 1599–621.
McMahon, R.J. "How Periphery Became the Centre: The Cold War, the Third World, and the Transformation in US Strategic Thinking." In *Foreign Policy at the Periphery: The Shifting Margins of US International Relations Since World*

War II, edited by B. Sewell and M. Ryan, 19–35. Lexington: University Press of Kentucky, 2017.

Mearsheimer, J.J. *The Tragedy of Great Power Politics*. New York: Norton & Company, 2014.

Mérand, F., ed. *Coping with Geopolitical Decline: The United States in European Perspective*. Montreal: McGill-Queen's University Press, 2020.

Mercer, J. "Rationality and Psychology in International Politics." *International Organization* 59, no. 1 (2005): 77–106.

Middlebrook, M. *The Falklands War*. Barnsley: Pen & Sword Military, 2014.

– *Operation Corporate: The Story of the Falklands War, 1982*. London: Viking, 1985.

Modelski, G., and P. Morgan. "Understanding Global War." *Journal of Conflict Resolution* 29, no. 3 (1985): 391–417.

Moffit, B. *The Global Rise of Populism: Performance, Political Style, and Representation*. Stanford: Stanford University Press, 2016.

Moore, C. *Margaret Thatcher: The Authorized Biography: Volume One*. London: Penguin Books, 2014.

Morgan, C., and C. Anderson. "Democratic Support and Diversionary External Conflict in Great Britain 1950–1992." *Journal of Politics* 6, no. 3 (1999): 799–814.

Morris, J. "How Great Is Britain? Powers, Responsibility and Britain's Future Global Role." *British Journal of Politics and International Relations* 13 (2011): 326–47.

Mosse, G.L. *The Image of Man: The Creation of Modern Masculinity*. Oxford: Oxford University Press, 1996.

Mueller, J.E. *Retreat from Doomsday: The Obsolescence of Major War*. New York: Masic Books, 1989.

Murray, M. *The Structure for Recognition in International Relations: Status, Recognition, and Rising Powers*. Oxford: Oxford University Press, 2018.

– "The Struggle for Recognition in International Politics: Security, Identity and the Quest for Power." PhD dissertation, University of Chicago, 2008.

Navarro, P. *Crouching Tiger: What China's Militarism Means for the World*. Buffalo: Prometheus Books, 2015.

Nolan, C. "Great Powers and International Society." In *The Empire of Security and the Safety of the People*, edited by W. Bain, 71–93. London: Routledge, 2006.

Norpoth, H. "The Falklands War and Government Popularity in Britain: Rally without Conquer or Surge without Decline?" *Electoral Studies* 6, no. 1 (1987): 3–16.

Nussbaum, M.C. *Upheavals of Thought: The Intelligence of Emotions*. Cambridge: Cambridge University Press, 2001.

Oakes, A. "Diversionary War and Argentina's Invasion of the Falkland Islands." *Security Studies* 15, no. 3 (2006): 431–63.

O'Dowd, E.C. *Military Strategy in the Third Indochina War: The Last Maoist War*. Heboker: Routledge, 2007.

O'Neill, B. *Honor, Symbols, and War*. Ann Arbor: University of Michigan Press, 2001.

Onea, T.A. "Between Dominance and Decline: Status Anxiety and Great Power Rivalry." *Review of International Studies* 40 (2014): 125–52.

O'Shaughnessy, H. *Grenada: An Eyewitness Account of the US Invasion and the Caribbean History That Provoked It*. London: Hamisch and Hamilton Ltd., 1984.

– *Grenada: Revolution, Invasion and Aftermath*. London: Hamisch Hamilton Ltd., 1984.

Paul, T.V. *Asymmetric Conflicts: War Initiation by Weaker Powers*. Cambridge: Cambridge University Press, 1994.

Paul, T.V., D. Welch Larson, and W.C. Wolhforth, eds. *Status in World Politics*. Cambridge: Cambridge University Press, 2014.

Parashar, S. "Anger, War and Feminist Storytelling." In *Politics and War*, edited by L. Ahäll and T. Gregory, 71–85. London: Routledge, 2015.

Pardesi, M.S. "Is India a Great Power? Understanding Great Power Status in Contemporary International Relations." *Asian Security* 11, no. 1 (2015): 130.

Parr, H. "Representations of Grief and the Falklands War." In *Politics and War*, edited by L. Ahäll and T. Gregory, 154–66. London: Routledge, 2015.

Paxton, R.O., and J. Hessler. *L'Europe au XXème siècle*. Paris : Editions Talladier, 2011.

Payne, A., P. Sutton, and T. Thorndike. *Grenada Revolution and Invasion*. London: Palgrave Macmillan, 1984.

Peterson, C.C. "Soviet Military Objectives in the Arctic Theater." *Naval War College Review* 40, no. 4 (1987): 3–22.

Petiteville, F., and D. Placidi-Frot, eds. *Négociations internationales*. Paris : Sciencespo Les Presses, 2013.

Perez, S. *Le Corps du Roi: Incarner l'Etat de Philippe Auguste à Louis-Philippe*. Paris : Perrin, 2018.

Perl, R., and E.E. Larson. *The Falklands Dispute in International Law and Politics: A Documentary Sourcebook*. Sydney: Oceana Publications, 1983.

Plourde, A.M. "The Origins of Prestige, Goods as Honest Signals of Skills and Knowledge." *Human Nature* 19, no. 4 (2008): 374–88.

Ponton, D.M. "The Female Political Leader: A Study of Gender-Identity in the Case of Margaret Thatcher." *Journal of Language and Politics* 9, no. 2 (2010): 195–218.

Popper, K. *Conjonctures and Refutationz: The Growth of Scientific Knowledge*, London: Routledge, 2002 (1963).

Post, J., ed. *The Psychological Assessment of Political Leaders: With Profiles of Saddam Hussein and Bill Clinton*. Ann Arbor: University of Michigan Press, 2003.

Pottier, P. "A French Experience in Counterinsurgency during the French Indochina War." *Small Wars and Insurgencies* 16, no. 2 (2007): 125–46.

Pu, X., and R.L. Schweller. "After Unipolarity: China's Visions of International Order in an Era of US Decline." *International Security* 36, no. 1 (2011): 41–72.

Przeworksi, A., M.E. Alvarez, A. Cheibu, and F. Limongi. *Democracy and Development: Political Institutions and Well-Being in the World, 1950–1990*. Cambridge: Cambridge University Press, 2000.

Raine, S., and C. Le Mière. "Southeast Asia: Between Emerging Great Power Rivalry." *Adelphi Series* 52, nos 436–7 (2013): 105–50.

Ramel, F., and J-V. Holeindre, eds. *La fin des guerres majeures?* Paris : Economica, 2010.

Reagan, R. *An American Life: The Autobiography*. New York: Simon and Shuster, 1990.

– *The Reagan Diaries*. Edited by Douglas Brinkley. New York: Harper Perennial, 2007.

Reinert, M. "Mondes lexicaux stabilisés et analyse statistique de discours." JADT, 9èmes Journées Internationales d'Analyse Statistique des Données Textuelles. Lyon 12–14 March 2008.

Renshon, J. *Fighting for Status: Hierarchy and Conflict in World Politics*. Princeton: Princeton University Press, 2017.

Reus-Smit, C., and D. Snidal. "Reuniting Ethics and Social Science: The Oxford Handbook of International Relations." *Ethics and International Affairs* 22, no. 3 (2008): 261–71.

Reynaud, A. *Société, espace et justice: Inégalités régionales et justice socio-spatiale*. Paris : P.U.F., 1981.

Richards, D., T.C. Morgan, R.K. Wilson, V.L. Schwebach, and G.D. Young. "Good Times, Bad Times, and the Diversionary Use of Force." *Journal of Conflict Resolution* 37, no. 3 (1993): 504–36.

Ringmar, E., and T. Lindemann, eds. *The International Politics of Recognition*. Boulder, CO: Paradigm Publishers, 2014.

Robbins, K. *The Eclipse of a Great Power: Modern Britain 1870–1992*. London: Longman, 1994.

Rose, G. "Neo-Classical Realism and Theories of Foreign Policy." *World Politics* 51, no. 1 (1998): 144–72.

Rotberg, R.I., and T.K. Rabb, eds. *The Origins and Prevention of Major Wars*. Cambridge: Cambridge University Press, 1984.

Rubner, M. "The Reagan Administration, the 1973 War Powers Resolution, and the Invasion of Grenada." *Political Science Quarterly* 100, no. 4 (1986): 627–47.

Ruscio, A. "L'opinion française et la guerre d'Indochine (1945–1954)." Sondages et témoignages. *Vingtième siècle revue d'Histoire* 29 (1991): 35–46.

– *La Guerre Française d'Indochine*. Bruxelles : Editions Complexe, 1992.

Salter, G. *Carl Schmitt: Law as Politics, Ideology and Strategic Myth*. London: Routledge, 2012.

Sanders, D., H. Ward, and T. Fletcher. "Government Popularity and the Falklands War: A Reassessment." *British Journal of Political Science* 17, no. 3 (1987): 281–313.

Sanders, D. *Losing an Empire, Finding a Role: British Foreign Policy Since 1945*. London: Macmillan, 1990.

Saurette, P. "You Dissin Me? Humiliations and Post 9/11 Global Politics." *Review of International Studies* 32, no. 3 (2006): 495–522.

Saurette, P., and K. Gordon. *The Changing Voice of the Anti-Abortion Movement: The Rise of "Pro-Woman" Rhetoric in Canada and the United States*. Toronto: University of Toronto Press, 2015.

Scott, H.M. *The Birth of a Great Power System 1740–1815*. Harlow: Pearson Education Ltd., 2006.

Seabury, P., and W.A. McDougall, eds. *The Grenada Papers: The Inside Story of the Grenadian Revolution and the Making of a Totalitarian State – As Told in Captured Documents*. San Francisco: Institute for Contemporary Studies, 1984.

Schenoni, L., S. Braniff, and J. Battaglino. "Was the Malvinas/Falklands a Diversionary War? A Prospect Theory Reinterpretation of Argentina's Decline." *Security Studies* 29, no. 1 (2020): 34–63.

Schmitt, M.N., and L.C. Green. "The Falklands Crisis and the Laws of War." *International Law Studies* 70 (1985): 203–13.

Schmitt, O. "Decline in Denial: France after 1945." In *Coping with Geopolitical Decline: The United States in European Perspective*, edited by F. Mérand, 107–26. Montreal: McGill-Queen's University Press, 2020.

Schweizer, P. *Reagan's War: The Epic Story of His Forty-Year Struggle and Final Triumph over Communism*. New York: Doubleday, 2002.

Schweller, R.L. *Deadly Imbalances: Tripolarity and Hitler's Strategy of World Conquest*. New York: Columbia University Press, 1998.

Sewell, B., and M. Ryan. *Foreign Policy at the Periphery: The Shifting Margins of US International Relations Since World War II*. Lexington: University Press of Kentucky, 2017.

Sharp, P. *Thatcher's Diplomacy: The Revival of British Foreign Policy*. London: Palgrave Macmillan, 2017.

Shepherd, L.J., ed. *Gender Matters in Global Politics: A Feminist Introduction to International Relations*. New York: Routledge, 2010.

Shibles, W. "Feminism and the Cognitive Theory of Emotions." *Woman and Health* 17, no. 1 (1991): 57–69.

Shifrinson, J.I. *Falling Giants: How Great Powers Exploit Power Shifts*. Ithaca: Cornell University Press, 2018.

Sjoberg, L. *Gendering Global Conflict: Toward a Feminist Theory of War*. New York: Columbia University Press, 2013.

– *Gender, War, and Conflict*. Cambridge: Polity Press, 2014.

Smith, A. "International Crisis and Domestic Politics." *American Political Science Review* 92, no. 3 (1998): 623–38.

Snyder, J. *Myths of Empire: Domestic Politics and International Ambition*. Ithaca: Cornell University Press, 1991.

Sobek, D. "Rallying around the Podesta: Testing Diversionary Theory across Time." *Journal of Peace Research* 44, no. 1 (2007): 29–45.

Solomon, T. "Embodiment, Emotions, and Materialism in International Politics." In *Emotions, Politics and War*, edited by L. Ahäll and T. Gregory, 58–70. London: Routledge, 2015.

Sontag, S. *On Photography*. New York: Picador, 1977.

– *Regarding the Pain of Others*. London: Penguin, 2003.

Soutou, G.H. *La Guerre de 50 ans: Les Relations Est-Ouest 1943–1990*. Paris : Arthème Fayard, 2004.

Specter, M. "Grossraum and Geopolitics: Resituating Schmitt in an Atlantic Context." *History and Theory* 56, no. 3 (2016): 398–406.

Sperling, V. "Putin's Macho Personality Cult." *Communist and Post-Communist Studies* 40 (2016): 13–23.

Stets, J.E, and P.J. Burke. "Femininity/Masculinity." In *Encyclopedia of Sociology*, (revised edition) edited by E. Borgatta and R. Montgomery, 997–1005. New York: Macmillan, 2000.

Suzuki, S. "Seeking 'Legitimate' Great Power Status in Post-Cold War International Society: China's and Japan's Participation in UNPKO." *International Relations* 22, no. 1 (2008): 45–63.

Taliaferro, J.W. "Power Politics and the Balance of Risk: Hypotheses on Great Power Intervention in the Periphery." *Political Psychology* 25, no. 2 (2004): 117–211.

Tang, S. "Reputation, Cult of Reputation, and International Conflict." *Security Studies* 14, no. 1 (2005): 34–62.

Thatcher, M. *The Downing Street Years*. New York: Harper Collins, 1993.

Tir, J. "Territorial Diversion: Diversionary Theory of War and Territorial Dispute." *Journal of Politics* 72, no. 2 (2009): 413–25.

Tir, J., and M. Jasinski. "Domestic-Level Diversionary Theory of War: Targeting Minorities." *Journal of Conflict Resolution* 52, no. 5 (2008): 641–64.

Train, H. "An Analysis of the Falklands/Malvinas Islands Campaign." *Naval War College Review* 41, no. 1 (1988): 33–50.

Tucker, R. *The Purposes of American Power: An Essay on National Security*. New York: Praeger, 1981.

Vacs, A. "A Delicate Balance: Confrontation and Cooperation between Argentina and the United States in the 1980s." *Journal of Interamerican Studies in World Affairs* 31, no. 4 (1989): 23–60.

Vallaud, P. *La Guerre au XXIème Siècle*. Paris : Perrin, 2014.

Valette, J. *La Guerre d'Indochine 1945–1954*. Paris : Armand Colin, 1994.

Van Creveld, M. "The Waning of Major War." In *The Waning of Major War: Theories and Debates*, edited by R. Väyrynen, 97–112. London: Routledge, 2006.

Vayne, P. "Panem et circenses : l'évergétisme devant les sciences humaines." *Annales Economies, sociétés, civilisations* 24, no. 3 (1969): 785–825.

Volgy, T.J., R. Corbetta, K.A. Grant, and R.G. Baird. *Major Powers and the Quest for Status in International Politics: Global and Regional Perspectives*. New York: Palgrave Macmillan, 2011.

Wainstock, D., and R.L. Miller. *Indochine et Vietnam. 35 années de guerre : 1940–1975*. Paris : Nouveau Monde Editions, 2019.

Wallace, M.D. *War and Rank among Nations*. Lanham: Lexington Books, 1973.

Wallerstein, I. *The Modern World-System 1. Capitalist Agriculture and the Origins of the European Economy in the Sixteenth Century*. New York: Academia Press, 1974.

Walt, S. *The Origins of Alliances*. Ithaca: Cornell University Press, 1987.

– *Revolution and War*. Ithaca: Cornell University Press, 1996.

Waltz, K. *Man, State and War*. New York: Columbia University Press, 1959.

– *Theory of International Politics*. New York: McGraw-Hill, 1979.

Ward, S. *Status and the Challenge of Rising Powers*. Cambridge: Cambridge University Press, 2017.

Weber, C. *Faking It: US Hegemony in a "Post-Phallic" Era*. Minneapolis: University of Minnesota Press, 1999.

– *Simulating Sovereignty: Intervention, the State and Symbolic Exchange*. Cambridge: Cambridge University Press, 1995.

Weisiger, A. *Logics of War*. Cornell: Cornell University Press, 2013.

Welch, D.A. *Justice and the Genesis of War*. Cambridge: Cambridge University Press, 1993.

– *Painful Choices: A Theory of Foreign Policy Change*. Princeton: Princeton University Press, 2005.

– "Remember the Falklands?: Missed Lessons of a Misunderstood War." *International Journal* 52, no. 3 (1997): 483–507.

Welch, P. *US-Caribbean Relations: The Grenada Invasion and Caribbean Political Decision-Making*. Saarbrûcken: VDM Verlag Dr Müller, 2009.

Wheeler, L.K. "The Grenada Invasion: Expanding the Scope of International Humanitarian Intervention." *Boston College International and Comparative Law Review* 8, no. 2 (1985): 413–30.

Williams, D. "Grenada Five Years Later: An Investigative Report of the Grenada Foundation Inc. on the Consequences of the US-led 1983 Invasion of Grenada." Grenada: Family Records Centre, 1988.

Williams, P. "Miscalculation, Crisis Management and the Falklands Conflict." *World Today* 39, no. 4 (1983): 144–9.

Windsor, P. "Diplomatic Dimensions of the Falklands Crisis." *Journal of International Relations* 12, no. 1 (1983): 86–96.

Wodak, R., and M. Meyer, eds. *Methods of Critical Discourse Studies*. London: Sage, 2016.

Wolf, R. "Respect and Disrespect in International Politics: The Significance of Status Recognition." *International Theory* 3, no. 1 (2011): 105–42.

Wohlforth, W.C. "Honor as Interest in Russian Decision for War, 1600 1995." In *Honor among Nations: Intangible Interests and Foreign Policy*, edited by E. Abrams, 21–45. Washington, DC: Ethics and Public Policy Center, 1998.

– "Unipolarity, Status Competition, and Great Power War." *World Politics* 61, no. 1 (2009): 28–57.

Wood, E. "Performing Memory: Vladimir Putin and the Celebration of World War II in Russia." *Soviet and Post-Soviet Review* 38 (2011): 172–200.

Woodward, S. *One Hundred Days: The Memoirs of the Falkland Battle Group Commander*. Annapolis: Naval Institute Press, 1992.

Yahri-Milo, K. *Who Fights for Reputation: The Psychology of Leaders in International Conflict*. Princeton: Princeton University Press, 2018.

Zakaria, F. *From Wealth to Power: The Unusual Origins of America's World Role*. Princeton: Princeton University Press, 1998.

Zuiderhoek, A. *The Politics of Munificence in the Roman Empire: Citizens, Elites and Benefactors in Asia Minor*. Cambridge: Cambridge University Press, 2009.

INDEX